ORIGINS

Origins

The
Cosmos
in
Verse

Joseph Conlon

A ONEWORLD BOOK

First published by Oneworld Publications Ltd, 2024

Copyright © Joseph Conlon, 2024

The moral right of Joseph Conlon to be identified as the Author of this work has been asserted by him in accordance with the Copyright, Designs, and Patents Act 1988.

All rights reserved
Copyright under Berne Convention
A CIP record for this title is available from the British Library

ISBN 978-0-86154-911-5
eISBN 978-0-86154-912-2

Designed and typeset by Tetragon, London
Printed and bound in Great Britain by Clays Ltd, Elcograf S.p.A.

Oneworld Publications Ltd
10 Bloomsbury Street
London WC1B 3SR
England

Stay up to date with the latest books, special offers, and exclusive content from Oneworld with our newsletter

Sign up on our website
oneworld-publications.com

Contents

Preface to *Elements*	9
Elements	11
Preface to *Galaxies*	75
Galaxies	77
Notes on *Elements*	127
Notes on *Galaxies*	147
ACKNOWLEDGEMENTS	159

Elements

Preface to *Elements*

This poem describes the formation and first origin of the elements of the Periodic Table. These elements had their birth in three separate epochs, which also define the three main parts of the poem.

The first epoch belongs to the very early universe, in the Hot Big Bang during the initial few seconds of the universe's existence. During this period, Hydrogen, Helium and a small amount of Lithium were formed.

The second epoch starts around a hundred million years later as stars begin to form. During the stellar life cycle, all the elements up to and including Iron are made through nuclear fusion.

The third period of original element formation occurs as heavy stars reach the ends of their lives. As these heavy stars run out of fuel, they collapse and explode as supernovae. During the apocalyptic environment in the immediate aftermath of stellar explosions, the remaining heavy elements, up to Uranium, can be formed.

The elements are crucial to life; almost all the elements in our bodies were, at some point, cataclysmically ejected from a star.

As well as for emphasis, italics are also used to indicate where a 'normal' word has a technical meaning, e.g. 'glue' or 'up'.

Elements

Before the silver clink and tortured 'Amen'
Of Jesus kissed by the Iscariot,
Before the worldly wealth of Tutankhamen
Was buried with his golden chariot,

Before the bronze of Theseus's blade
Had drained the lifeblood from the Minotaur,
Before a node of Carbon cooled and made
A smudge of pressured muck the Koh-i-Noor,

There raged a thumping cosmic ballyhoo,
A manic dance – a rumpus to arouse
The universe: of Higgs and W,
Electrons, gluons, muons, Zs and taus,

Plus three neutrini and six types of quark.
One part invisible, two parts unstable –
What led this turbulent primaeval spark
To populate the Periodic Table?

ORIGINS

O rows and columns! Dense with cryptic print,
As in some diabolic crossword game,
Your letters, numbers — numbered letters — hint
That hidden logics lurk. The marks proclaim

Structure to your exotic toxic lists,
Perfect for poisonings and pyrotechnics.
What are these elements? The alchemists,
Dying to know, disdaining dialectics,

Once waged with fire and acid their campaign
To turn a common metal into gold
And so get filthy rich. They died in vain —
Their quest was doomed by ignorance. To mould

An element takes more than human hands
And needs extreme high-temperature conditions
Not found in home-brew labs. To understand
Why this is so, and why nuclear transitions

Were their unknown goal, we must comprehend
The atom, gross details and inner nature.
Basics come first; the wise (and I) commend
Command of concepts over nomenclature.

ELEMENTS

All atoms have two parts: an outer cloud
And central nucleus. The cloud consists
Of loose electrons, bound to paths allowed
By quantum theory. Mutual force exists

Between these two, set by electric charge:
Though each electron counts just minus one,
The corresponding nuclear term is large
And positive. From this atomic Sun

Extend out tentacles of force, like blind
Attractive sirens. Stylish algebra
Shows these permit the nucleus to bind
Its satellite electrons – as our star,

Compact and central, leashes Uranus
Two billion miles away. The physics rhymes –
Electrons' orbits round the nucleus
Surpass its size one hundred thousand times.

Electrons, we believe, are fundamental:
On current measurement, a fair deduction.
The nucleus, for sure, is accidental
And so amenable to deconstruction.

ORIGINS

So break its tiny Lego up to find
Two types of piece, and each attached by *glue*.
(But not the Airfix sort, or other kinds
Of sticky-fingered mess well known to you

From cardboard crowns or sprigs of model gorse.
This is not squeezed from tubes of gunk with screw-on
Finicky lids. This is the nuclear force,
With etymology 'derived from gluon'.)

What are these blocks? The first, and fountainhead,
Are protons. Charged, they fix the harmonies
Inside each atom and are ballasted
With neutral neutrons (hence the name). Through these

Constituents, all nuclei are made,
In builds mere femtometres in their size.
One page of verse! – which Newton would have paid
In golden crowns to read, and realise:

What makes an element? The summed amount
Of protons present. What are isotopes?
Fix proton number; vary neutron count
And climb the sheer radioactive slopes

ELEMENTS

Which flank the valley of stability
Where dwell the elements we know from Earth,
The first and oldest nuclear family.
Now hear of each, its heraldry and birth.

We start with Hydrogen, both first to form
And also simplest, for a single proton
Defines its character. Unlike the norm,
It has no neutrons. Still, a little note on

Internal structure: this consists of *quarks*,
A shackled triplet written *uud*,
Two *ups* and then one *down* (convention marks
Them by initials) joined in harmony

To make one object. Tethered to and with
Their fellow quarks, this triune synergy
Defines the proton, whose externals live
Above a foaming inner *quantum sea*

ORIGINS

Of *virtual* quark-infested *strong-force* slop,
Where heavy *bottoms* and their anti-twins
Play quantum peek-a-boo. Without a *top*,
A state less *charmed* than *strange*, its outs and ins

Arise from swarms of bound and coupled pairs,
Down and then *up* in manifold positions.
Now here, now there: these intricate affairs
Provoked and urged subnuclear theoreticians

To probe its innards through the residue
Left by colliding bodies in reaction.
Along with quarks, here they detected *glue* –
Especially in the *hardest interactions*.

But, from this quantum swirl of *legs and loops*
We shall avert our eyes. Their moxie risen,
New couplings formed – yet still the proton coops
Them up, confined within its quantum prison

To seethe in vain, enslaved and trapped inside.
Outside its walls, no single quarks survive;
Once protons settle, we are justified
To cast to oubliettes their inner lives.

ELEMENTS

But how did protons form? From cooling soup
Of quarks and gluons — mixtures then so hot
That any friendly quarks which tried to group
Themselves to make a proton would be shot

Apart like th---------is, their bonds dissolved by heat.
For heat is energy, and fragile links
Can only take so much and stay complete.
Like some provincial Juliet, who thinks

The world of boys is bounded by her school
And centred on her one true love — then goes
Away to college, sees him for a fool
And falls for dashing worldly Romeos.

And as the cosmos grows, it also cools —
So proto-protons are bombarded less.
Heat which expunges would-be molecules
Now merely makes new protons effervesce.

Then from this gloop of X-rayted miasma
Fall tiny globules tethered tight together,
Which turn the cosmic soup into a plasma
Of particles which, ever after, never

ORIGINS

Again will split apart to quark and gluon –
Though from the older forms are born their scions
Which spice the broth of positron and muon
With stronger flavours: kaons, neutrons, pions

And protons. Proton: what's behind that name?
Science may stay indifferent – if not men and
Women – when different is declared the same,
Like protons and ionic Hydrogen: and

Truth marches on. Can this be progress? Can it?
They found out much, advancing knowledge far,
Who grasped that Venus, though a single planet,
Is both the Morning and the Evening Star.

Progress consists of more than adding more
To lists of who found what. More useful acts
May lie in posing probing questions or
Restructuring the known existing facts –

And facts and particles were both restructured
When quarks all merged and stuck, and then stayed stuck,
To leave the older dispensation ruptured,
Where quarks were three, and freely charged amuck.

ELEMENTS

That past is lost; its passing compensates
Through novelties thus brought to life. For when
The quarks no longer lived as stable states,
A new and fresh epoch commenced. And then,

That happy moment, ever-elemental,
Always-plentiful Hydrogen was made –
The act foundational and fundamental
To all of chemistry and all its trade.

First-born and first in numbering, this mite
Does little by itself. In company,
It comes to life; an eager appetite
To saturate another's valency

Gives it the chemical creative tools
To build up meth- and eth- and all the -anes,
The most baroque organic molecules
And caterpillar hydrocarbon chains.

Not yet. To make a partner element
We need, as well as protons, neutrons too,
Which, by attraction or by accident,
Must stick to protons using strong-force *glue*.

ORIGINS

Like number one, the neutron's built from quarks,
A shackled triplet written *udd*,
One *up* and then two *downs* (convention marks
Them by initials) joined in harmony

To make one object. Tethered to and with
Their fellow quarks, this triune synergy
Defines the neutron, whose externals live
Above a foaming inner *quantum sea*

Whose image we have met. Except for charge,
The two are twins, with both their size and mass
Almost the same. But when these splash and sparge
On protons – the pervasive neutron gas

Could hardly miss – some stick and, staying bound,
Form a new composite, the *deuteron*,
Built from both p and n. This can be found
In D_2O, a doppelgänger on

Aquatic roads. They look so much alike,
But looks deceive. Go drink a bit – to feel
No change at all. Go drink a lot: it strikes,
A knife disguised as water, to reveal

ELEMENTS

A poison of the flesh it irrigates.
Though barely stable, deuterons survive
Until they too are hit, when infiltrates
One further neutron, whose impact contrives

To form some *tritium*, or *pnn* —
The rarest, heaviest, and third and final,
Natural isotope of Hydrogen.
In making Helium, these three combine, all

Fusing with each other. For this to happen,
They first must close and touch. Then, having touched,
The strong force swoops and seizes them, to trap in
Its intra-nuclear web. There, partly clutched,

They squirm and wriggle and, though some escape,
Many are held. These hostages release
Their surplus energy, then find the shape
Preferred by physics, marked by much increased

Stability. This, in its settled state,
Involves a four-hand dance of two and two,
Neutron and proton pairs. These circulate
Around themselves. Like angels, they eschew

ORIGINS

External stimuli or time's decays.
For this is Helium, the Nobel gas,
Inert and stable. Born, it lives and stays
How it was. Ages come; more ages pass;

Its dance goes on. Before the oldest stars,
It was. And when the stars have lost their lights,
It shall still be. Were it to write memoirs,
They would be deathly. 'Through unending nights,

I was, and nothing changed. Now I am here.'
Almost all Big Bang Helium produced
Remains intact, the fraction lost a mere
Grain to a granary. Excitement-proof,

It staggers on. What makes this bore so stable?
In nuclei, the strong force waxes rude
And plain electric fields are quite unable
To match its pull: a sardined multitude

Of self-repulsive plus-charged protons may
Be battened down. Their strong bonds dominate –
O queen of forces! – and these chains outweigh
Electric tendencies to separate

ELEMENTS

And scatter protons, out far and alone.
Such are the reasons why, once they are made,
The nuclei survive. Around this throne,
The atom's central fortified stockade,

The strong force rules supreme. Once in its claws,
All older liberties must be forgot.
The thrall is small but total; here its laws
Dictate what may survive and what may not.

Like great Goliath's pecs of burnished steel,
The price for brutish strength is lack of range.
Although up close all others have to kneel,
On greater distances the standings change:

And though large elements, like Iron and Gold,
Will last forever, crafted in their proper
And stable form (such features also hold
For Silver, Sulphur, Chromium and Copper)

To reach this place is hard. Far off and masked,
All these alluring nuclear auras fade,
Leaving one awkward question to be asked:
How can the larger nuclei be made?

ORIGINS

These only form if protons meet and merge
But protons carry an electric charge,
Charges which interact as they converge
And near convergence, these effects grow large.

Like shepherd David's polished stones and sling,
Electric forces zap out from afar.
Villeins up close, distance anoints them king;
Last is now first; the kulak crowned as tsar.

But charge repels itself — this simple fact
Demands fizz-popping whizz and bang to broach
The wall of force that nature will enact
To bar protonic twins from close approach

(For laws which underpin electrostatics
And sharpening, with insight and aplomb,
Practical rules of thumb to mathematics,
I raise a glass or six to Charles Coulomb.

A modern corps of army engineers
May make great walls restraining soil and sea,
Or build imposing forts and crested weirs
With cheerful humour and alacrity.

ELEMENTS

But they would not contain among their ranks
One such as he, who added to this load
Floods of discoveries which broke the banks
In which the streams of human knowledge flowed) –

That every time two protons halve their distance,
The force quadruples to oppose their course,
Requiring bumptious spirit or assistance
To break down these electric walls of force.

No subtleties; no fancy stratagem.
To drive the protons down to these regimes,
Pack them with energy, then up and at 'em
With oomph to plough straight through – like rugby teams

Whose backs dart forth in vain into the teeth
Of mauling tackle packs, so toss the ball
To twenty ugly stone of tighthead beef
Steaming and barrelling towards that wall.

Where artfulness fails, impetus succeeds.
The charging tank smacks down the flailing hosts,
Who cling to arms and legs with futile deeds –
He grounds the ball and them beneath the posts.

ORIGINS

And in cosmology, the Hot Big Bang
Weaned all its particles with energies,
With vim and vigour needed to defang
The nuclear Cerberus. All those degrees

Create abundant heat, and heat is work:
And work enough to drag each inbound pair
Through every last repulsive bump and jerk
Caused by their charge. They touch; they kiss – and there

They merge to Helium. To make the rest,
Redo this all at scale. Two nuclei
Approach at high momenta, climb and crest
The steep electric slopes, then unify,

Fused into one. The product nucleus,
Enriched with protons, puffs up with increased
Atomic number. Welding ion plus
Ion thus builds the elements, from least

To greatest. Not too fast – for this ascent
Proceeds through gradual steps, which often stop
At midway staging points to circumvent
Impossibly direct routes to the top.

ELEMENTS

When Helium absorbs one proton in,
The merger leads to Lithium. If it
Instead collides and fuses with its twin
The outcome is Beryllium. Admit

One further Helium – we now have Carbon;
An extra proton makes that Nitrogen,
Then move through Sulphur, Phosphorus and Argon
To Iron, and all up-built from Hydrogen!

Not yet – the universe, now aged one second,
Seethes with its youthful zest. It roughly shakes
Its raw ingredients, as if to check on
Their quality. The act of mixing makes

This broth an *amuse-bouche* for chemistry:
A blend of Hydrogen and Helium
(Though early universe gastronomy
Adds in a soupçon's worth of Lithium.

Electrons stabilise the cosmographic
Chowder; the photons give it bite. One teeny
Last-least addition, near homeopathic
In flavouring: essence of three neutrini.)

But through each element, as charges grow,
Their broad electrified shield-wall sparks stronger.
To reach the guarded nuclear heart below
The inward path becomes harder and longer.

And, as we saw, a growing cosmos cools.
For particles which ride its wild affray,
Cosmic expansion claims its tax in joules
And elemental dreams dissolve away.

The Hot Big Bang is hot – but not enough.
It gives its all. The atoms charge at speed.
They close and nearly touch – then are rebuffed.
Knocked back to whence they came, the pairs recede.

Like modern Hercules, whose grunting heave
Reaches for glory in the clean and jerk –
His face contorts with strain; his muscles squeeze.
He screams. His tendons tauten with the work,

Then snap – and all he sacrificed, to peak
In an Olympic year, is lost. He calls on
His God in vain – torn tissues regrow weak;
The medal sensed, and partly grasped, now falls, gone.

ELEMENTS

The Hot Big Bang now drops subcritical;
Its youthful fervour chills and cools down fast.
As closing speeds become testudinal,
The time for scaling nuclear walls has passed:

And so concludes that far-off distant era
When elements first formed. It lasted mere
Minutes — ten madcap minutes that were queerer
Than any you or I have known. Its sphere

Of influence extends to now: for gas
Found in the deepest voids of space, when split
By Hydrogen and Helium, has mass
Ratios first established then. This bit

Of data, hard fact hard-earned, is one way
We know this ancient story's truth, this tale
Of times when time still lacked a name, of clay
To mould a universe — and if I fail

To show its majesty, and make it seem
Some dumb, fantastical charade, my fault
Pains not just me alone, but all who dream,
Or dare to dream, of truth. But now we halt,

ORIGINS

To think on where our story stands and where
It must go next. Despite this upbeat start,
The start stopped, with its residue the pair
Of Hydrogen and Helium. The part

That makes up Lithium is minimal,
With larger atoms not yet there at all.
But turn your thoughts to now: our world throbs full
Of elements, like some escutcheoned hall

Abuzz with noise, close-thronged with every nation,
Where wines and candles blaze alike, which thrums
To music, poetry and conversation.
But early on, some mendicant who comes

And knocks, in search of others' company
To break his heavy-footed lonely grind
And flee the crushing uniformity
Which fills the early universe, would find

The hall bereft: three guests, and only three,
Sat at the Table of the Elements.
All else is desolate – no Mercury,
No Tin. To overcome their reticence,

ELEMENTS

He waits —
    waits —
        waits —
            waits aeons, half eternity,
For stars to form, when in their furnace-heart
The fiery pressured forge of gravity
Can fuse those nuclei now held apart.

What makes a star? Infalling dust and gas
Piled up by gravitation's constant pull,
As specks of cosmic fluff and extra mass
Are gently gathered. Lake Windemere, full

Of water, brims with its uncounted drops
Of rain. Cathedrals grew up stone by stone,
While those who climb like goats to mountain tops
Ascend the crags with muscle strength alone.

ORIGINS

All these need time – as do our stars. The knife
Which slices thin through time again and again
Reveals but gradual differences – not strife,
Not clashing armies on a darkling plain,

But steady growth of slightly denser spots
Where chance's starting dice threw double-six.
These nodules slowly grow. The cosmic plots
Of mass blush deeper shades. Their hue depicts

The growth of structure, as the haves accrete
From have-not neighbours, and the matter-poor
Endure the principle found at the meet
Of biblical and scientific law:

'Let them that hath be given more.' Or so
Claimed the cosmologist of Galilee.
Smirk all you like – good theoreticians know
It's true in General Relativity.

What is a star? A balanced stable locus,
An equilibrium between two ends.
At one sits gravity, with all its focus
On straight collapse. It tugs and pulls. It bends

ELEMENTS

Trajectories to point inwards. It draws
Escaping gas back and inside. The goal:
Drive everything towards one point, to cause
The star's replacement by a new black hole.

Opposing this, an outward pressure blows
Up inner layers like inflated balls.
Alone, it would explode its host. It flows
Out from the centre to the edge, then falls

To zero on that surface. On each shell
Of gas, on scaly stellar onion skins,
Gravity pulls and pressure pushes. Well
They fight and long they argue. If one wins,

The star will die. Their near-eternal tug-of-war
(Ten billion years – though may be more or less –
Of rivals wrestling out a close-fought draw)
Governs all stellar twinkly-twinkliness.

This source of pressure is our source of light,
As heat in constant flux up from the core
Not just supports the star but makes it bright,
Radiant, provident and fertile; for

This source of light supplies our source of life,
A warmth to nurture and embrace our spinning
Globe-nest, and to all fauna, first midwife.
But in this end, we see our true beginning:

This source of life produces elements.
Its deep-down wells of heat do not come free,
As manna from divine omnipotence,
But fire through fusion, loosing energy

Obtained from merging pairs of nuclei.
Drawn in by gravity, their mutual speed
Increases. Falling more, they multiply
Their impact, garnering the punch they need

To force a hole in the electric wall.
And if their first attempt does not succeed,
They try again, when once again they fall,
Pulled back towards the stellar core. This deed

Is thus repeated many times – until,
At last, it happens. In their starry cage,
Confined by gravity, the protons will
Collide and circulate. The star may age,

ELEMENTS

But holds its protons trapped – which slowly burn
From Hydrogen to Helium. They meet
That small bit heavier; they merge – and turn
Their extra mass to energy and heat.

What constitutes a star? This little pair
Are everywhere. Although on Earth, the two
(Especially Helium) are rather rare,
They dominate the Sun – a fact known through

The doctoral thesis of Cecilia Payne
(-Gaposchkin later on, in married life).
Born at the time when well-off girls with brains
Sensed prospects more than that of future wife,

Aged four her Daddy drowned. Her nursing mother –
Alone, a widow and a foreigner,
Raised her, her baby sister and their brother
To cherish music, arts and literature

At home beneath the Chiltern hills. At twelve,
They moved to London where the schools were better.
St Mary's, Paddington, first let her delve
In scholarship — but still within the fetter

Of Classical Curricula. To learn
More science, she transferred to St Paul's Girls.
Its mission was to educate — and spurn
The homely syllabus of knits and purls

For one of practical laboratories.
In halls expressly built for thinking space,
With staff hand-picked for subject expertise,
It taught its charges that a woman's place

Was good as any man's. From bluestockings
Of corseted éclat, she learned to balance
Science with Plato, Mozart and such things
As interested this girl of many talents.

Her music teacher, Gustav Holst, thought she
Should not pursue the planets, but embrace
His art instead: a compliment that we
Are glad was spurned. Next came a scholar's place

ELEMENTS

At Newnham College, Cambridge, there to read
Natural Sciences (the course, but not
Degree – those were for men; that change would need
A further three decades). And here, the lot

Of future fortune fell her way. Her aim
Had been to specialise in botany.
One hero met, one talk – and she became
A debutante astronomer-to-be.

The hero, A. C. Seward, occupied
The Chair in Botany. Its contemplation
Had filled her puppy love; but he supplied
Neither encouragement nor inspiration.

But what to do instead? Chance commandeered
Her destiny. In vain, she'd hoped to sneak
A seat to hear the country's most revered
Learned astronomer, due soon to speak.

His talk was tightly ticketed; then fate
Made her friend ill and off she went (to see
Inside the college raised to celebrate
The Holy Undivided Trinity –

Or six-wives Henry's launderette for plunder).
As Eddington transfixed Great Hall with tests
Of general relativity, a thunder-
Clap rent her mind. Adrift, she could not rest

Or sleep for several days, then 'changed her shop'
And switched to physics, chief of sciences.
Though mostly liberated by the swap,
She was put off by Rutherford, with his

'*Ladies* and Gentlemen'. Such dashing wit!
The men all laughed, but she was not amused —
So visible, the rules required she sit
Alone in the front row. By night, she used

The Newnham telescope. She would extract
From star-filled skies two loves: a simple fun
In finding out; a joy in fresh-cut fact.
So she approached Professor Eddington

And asked to do research; her bosom full
Of hope, her head expecting brusque rejection.
'Hmmm. Well, my dear — a bit unusual.
Still, there is no insuperable objection

And I see nothing fatal with the notion.
Here – take these plates of Messier 36.
Go forth and calculate the proper motion.
Compare with theory; see what that predicts

And how the data match. If all goes well,
We'll publish it together.' Off she went,
Suffused with joy. A chat with Smart; a spell
Upon the measuring machine; she sent

For Gauss's works. 'I've finished. Here it is;
What do you think of it?' He read the text
And mailed it to the *Monthly Notices*,
Her first of many papers. What was next?

A schoolmarm's job. At least in England, this
Was all she could expect. But life was more
Than England; doors at home that prejudice
Or habit shut, elsewhere now quivered – for

The tide of change was coming in, the tide
Of post-war post-pandemic novelties:
The Charleston, flappers – all that jazz; a Slide
In Moral Standards; Ford and Model Ts.

A fellowship took her to Harvard; there
She joined the Cambridge women, pioneers
In stellar spectra. Free to ramble where
Her thoughts might lead, she had two blessed years

Of independence. Harvard had amassed
The largest set of stars yet known, a knot
Of curiosities whose reach surpassed
All other astral listings. But for what?

As tales to kindle distant unborn fires
Are found in novels not the dictionary,
So too with science; permanence requires
Theories which order mere spectroscopy.

Though spectra proved that distant stars possessed
Familiar elements, their ratios
Remained unclear. While expertise professed
That nothing new lay in the sun (which shows

What science thinks of expertise), there was
No actual evidence, just old heuristics
And prejudice. But not for long – because,
Combining quantum physics and statistics,

ELEMENTS

Saha of India had now derived
New formulae to fix the shapes and strengths
Of spectral lines – ideas, it seemed, contrived
Exquisitely for her to test. Those lengths

Of time in Cambridge libraries, those nights
Tucked up and snuggled with the stars: all went
In one unworldly cocktail. Eremites
Drop all for God; so scientists who scent

The wispy trace of undiscovered truth
Surrender everything to get there first.
She worked and worked. The energy of youth
Was rocket fuel; fortune reimbursed

Her tenfold for her toil. With cigarettes
For daily bread, she chimney-smoked her way
To understanding. Harvard's datasets
Gave her their secrets; she did not betray

Their trust. Most elements, as was suspected,
Came with abundances the same as Earth,
Or nearabout. But not all – she detected
Two fugitives way out of bounds. A dearth

ORIGINS

Of them was not the issue: Hydrogen
And Helium were both far too abundant,
The latter (think name) by one million,
The former, only thousandfold. Redundant

Theories, though, slowly die. Such ratios
Were staggering, the fruit of nights of bustle
And days engrossed in spectra — but not those
Preferred by Princeton's Henry Norris Russell.

He was not cross; he neither raved nor ranted.
But he was dominant and powerful,
And said they must be wrong — so she recanted
And said these numbers were unphysical.

Eppur si muove! Time taught him and us
That she was first and she was right; but she
Was still a she — and though no scholar does
Her work for riches, plaudits still would be

Welcome. When later, Russell changed his mind,
Though he acknowledged her, he got the credit
And all the approbation of mankind —
Until time's backwash thundered in, to edit,

ELEMENTS

Erode, reclaim, tear down old statuary
And there erect her lasting requiem:
The lustrous twinkling night-time potpourri
Is made from Hydrogen and Helium.

Of stars which punctuate the darkened sky,
Most shine like this. Unchanged from year to year,
They seem eternal – not like us who die,
In mortal trappings doomed to disappear.

But fires that lose their coal must dull and cool,
Their flames a parting gift from food that's gone.
The same for stars which live on nuclear fuel –
Time's footsteps tread wherever starlight shone.

Time's merciless and universal ratchet
Exhausts a star's reserves of Hydrogen
And, having slipped, no human act can snatch it
Back, or unwind its cogs to start again.

ORIGINS

As greater mass creates a greater pull,
Requiring greater pressure for support,
Big stars die first. By nature, bountiful
In burning fuel for heat, they soon run short

And then run out. But where these monsters lead,
The tiddlers follow, down this road which takes
More time or less, but always ends with greed
Unsatisfied and frenzied thirst which makes

The core contract — but, as this shrinks, it blows
Its aery outer layers up. Like pliant
Yet prickly buds unfolding to a rose,
These bloom through death-dark space as a red giant.

But in still-shrivelled depths, the wizened core,
Now made of Helium alone, sits stewing.
Its temperature ticks up, then rises more,
A rise that marks the behemoth's undoing.

Some Helium, spurred on by heat, slips past
Its own defences. There, it finds its twin
And forms Beryllium. This fails to last —
It briefly flits, but falls apart within

ELEMENTS

One femtosecond. But that single tick
Of tiny time allows one fleeting kiss
From passing Helium. Their natures click;
Out pops a Carbon resonance. If this

Form of excited Carbon, if this state
And mass did not exist – then nor would we.
Through its decays, couplings and just-right weight,
This Goldilocks, sublimely crafted key

Unlocks the gate which bars the road to life.
For when excited Carbon de-excites,
It turns to stable Carbon – which is rife
Wherever life has been: in trilobites,

In turtles, tortoises and dinosaurs,
In sneezewort, old man's beard and snake's head iris;
In goldfish, guinea pigs and Labradors;
In Made-In-China coronavirus.

This state was not identified by chance
But via logic out of Conan Doyle.
The Watson-Holmes duo of this advance:
Fowler of CalTech and Cambridge's Hoyle.

ORIGINS

'I am; you are; so Carbon once was made;
This only happens if this state exists.'
Through such anthropic flights did Hoyle persuade
The sober Fowler, nuclear physicist.

So Fowler went and searched, perhaps to find,
And find he did, exactly where Hoyle said –
Proof of the brilliant and fertile mind
In Fred's bespectacled abstracted head.

Joining the Burbidges, Margaret and Geoff,
They stressed in journals and in seminars
This fact whose truth will long survive their death:
Post-Carbon elements are born in stars.

The Nobel Hoyle deserved, he never won –
For Fred was Yorkshire born and Yorkshire blunt.
He spoke his mind; he was his county's son –
But damn-fool idiots would take affront.

Thinkers who chew the cud of others' thoughts
Grow old protected by the dundridge herd,
Where bovine sameness nurtures and supports
Those filled with fear of being found absurd.

ELEMENTS

But those original and great-souled minds
Who scorn received opinion with disdain
Die hard – their self-same independence blinds
Them to the fading strengths of ageing brains.

So Hoyle, consistent with his constitution,
Continued dreaming past his dream-by date,
Rejecting both Big Bang and evolution
For 'life from comets' and the Steady State.

But judge with gentleness, for know that he
Drank deep from truth before the final night
Closed this life bustling with desire to be
Provocative if wrong, not dull but right.

And through his Hoyle state, formed at resonance,
The stars make Carbon. This accumulates
Deep in their centres, but rich sustenance
This state is not – the core soon conflagrates

And flashes Helium to Carbon ash.
Even for common stars, the energies
Freed in this sudden intra-nuclear clash
Briefly outshine all normal galaxies.

What happens next depends on weight. All stars
Were born from clumping dust and gas; such clouds
Start their collapse at many sizes (ours
Lay in the middle). From these sparkling crowds

Filling the sky, of every mass and age
And size and origin, one plot is made
To trace their shared dynamics and so gauge
Both distant past and future. When displayed

By Russell and Hertzsprung, who graphed them all
By magnitude and visual shade, one line –
The tale of stellar evolution – sprawls
Clearly across this chart. Its ragged spline,

Plus calculation, shows where stars have been
And also, far from now, their future fate:
The laws of stellar astrophysics mean
That destiny is prophesied by weight.

Stars that stay small must end their lives as dwarfs,
Allowed to go off gently to the night –
As what shone brilliant so softly morphs
To nothingness, time dimming snow-white light.

ELEMENTS

But stars which more resemble Tweedledum,
And guzzle dust and gas and never stop,
Face ends befitting those who stuff their tum –
Fatties who overate, their tummies pop.

The lighter stars concern us less. They make,
Then halt at, Carbon. This survives, to stay
Trapped in the cooling core, but it would take
Catastrophe to haul it out. One way

That leads to this sometimes occurs in double
Stars, when one pulls its partner close and feeds it
Munchies to plump up dozy dwarfs. Thus trouble:
Like dogs, a dwarf's best left to lie; exceed its

Optimal weight and loose its wrath. At first,
All seems serene while Sleepy gorges. Then –
On judgement day – he starts, as though coerced,
To shrink in size but grow in mass, and when

This gets too much, it oversteps the weight
At which his core can hold him up. He turns
Inwards. The pressure starts to aggravate
His deep reserves of dormant rage. This burns

ORIGINS

Intensely up and out – where it creates
A supernova (type 1A). This bound
(That white dwarf stars will self-annihilate
At 1.4 our own sun's mass) was found

By youthful Subrahmanyan Chandrasekhar.
His boat from India was long and slow,
And no amount of puff-huff-puff could make her
Chuff quicker through the waves. But minds can go

Where steam cannot, to reach up through the smoke
And grasp the physics of exhausted stars.
These, having fired their stocks of nuclear coke,
Have nothing left to burn. Still, one thing bars

Such stars from straight collapse: Fermi repulsion,
Which holds electrons separate from each other.
This quantum form of pressure and revulsion
Lives on in stellar corpses. To uncover

Its physics was young Chandra's goal. He worked
At sea, he worked in Cambridge. Fluency
He earned through graft (life-long, he never shirked
The working habits set at twenty-three).

ELEMENTS

Opinion said, 'The quantum force will win,
Right at the last. This odd but truthful fact
Will hold stars up, prevent them falling in,
And leave their future cold, dark and compact.'

Opinion erred, through overly simplistic
Views on how quantum matter should behave.
Its starting point was not relativistic
And Chandra saw this imprecision gave

Mistakes which multiplied for massive stars.
The missing large-mass terms all serve to soften
The outward quantum pressure. *Ket* and *bra*
Were subtly altered; the presumptive coffin

Of all dead stars — a cold but self-supported
Black quantum dwarf — could only fit some part
Of them. His careful workings-out extorted
The last hurrah of massive stars — at heart

Self-gravitating balls of quantum gas.
Turn all the fusion off; completely dim it —
They *must* collapse above a certain mass
Much later named the Chandrasekhar limit.

Might he be wrong? The chance restrained his joy.
He thought it through again with trepidation,
Himself the critic scheming to destroy
His first-born child, this gorgeous calculation.

For such a major breakthrough, prudence owed it
To triple-check the minus signs. But, bless,
They held. He proudly-tentatively showed it
Before a meeting of the RAS.

Leading astronomers across the land
Had heard the news. Darkness's denizens,
Like black-tied crows they flocked to understand
Whether the laws for massive foreign suns

Had all been reconfigured by this son
Of foreign climes. Of views they coveted,
The chief was that of Arthur Eddington,
Knight of the realm and FRS. He said,

'Well, boy! I grant you it's a fair attempt
For a jejune apprentice of his trade.
But, to be frank, young man: not to pre-empt
The peer review, but look — don't be dismayed —

It makes no sense. You must have erred. Trust me —
I know. Your theories, sir, are quite absurd.'
Silence. A nervous cough. A break for tea
And sandwiches. 'Quite wet today.' He heard

The absent talk of other men. They sensed
His shame; they would be kind — but not agree
To intervene, still less to go against
The alpha monarch of astronomy.

Now, was this old man's pride? *Or was it race?*
Or both — who knows? What does it change? Abashed,
Chandra perceived his sudden loss of face
And public standing so abruptly dashed.

The flower that bares its first auspicious bloom
Into an unexpected springtime frost
May wither inside and its form assume
The might-have-been of promised beauty lost.

But not with Chandra — whose internal mettle,
Forged in this harsh put-down, cast on its own,
Revealed itself to be no fragile petal,
But tempered steel. In that dismissive tone,

ORIGINS

He heard the stale Imperial zeitgeist
And, though engrossed in esoterica,
He saw where future dawned. One look sufficed
To draw his spirit west. America,

Once land of pilgrims, knew its growing might
And sought out brains to match. It called across
The seas. Hands out, the New World glistened bright
With hope. He went; its gain was Europe's loss.

As Empire ebbed, he settled in Chicago,
The smartest of the smart. Kaleidoscopic
In range, he picked his subject. *Lento*, *largo*,
He mused – then wrote *the* book to parse that topic.

His honoured memory, his Nobel Prize,
We leave to rest in Illinois. His bound
Lives on. It takes his name forward and applies
Wherever stars are born – a breeding ground

That fills the whole known universe. The lesser
Stars, which stay light, all die as dwarfs in quiet
And dignified decline, still the possessor
Of all they made. The force-fed circus diet

ELEMENTS

Of supernova school is had by few.
Such cosmic fireworks, when they detonate,
Throw out and scatter elements. All true,
But rare – occurring only at a rate

Of one per galaxy per century.
The converse truth, of deep significance,
Is that the greater stars can never be
Themselves forever. In their countenance,

Which blazes fierce with the majestic red
Of ruby-rosebud onyx-specked attire
While X-ray necklaces highlight a head
Crowned with tiaras of magnetic fire,

We see death – and, along with death, discern
A voice: 'Remember, star, that thou wert dust
And unto dust and ash thou shalt return,
And there is no escape.' And so it must,

And so too must we. But, before the end,
How it creates! First Helium – we saw –
Is turned to Carbon through a pressured blend
Of nuclei. In lighter stars, no more

Than this occurs. But massive astral giants
See heat and pressure rise to the extent
They fuse these Carbon nuclei. The science
Of stellar structure and development

Reveals the final steps these heavies take.
When Carbon burns, its ash re-forms as Neon,
Magnesium and Sodium. The brakes
Come off; while Hydrogen burned for an aeon,

Its empire measured out in million-years,
This final stretch runs at a sprint. The stock
Of Carbon dwindles down, then disappears
In eye-blink centuries. The doomsday clock

Stands at eleven fifty-nine. It ticks
Greedily, keen to strike — but not just yet.
Carbon exhausted, now the nuclear mix
Demands that Neon underwrite the debt

Required to keep on keeping on. This pays
Its sorry way a year or two, this poor
Low-yielding chaff, a titbit which delays,
Not halts, oncoming Armageddon. Core

ELEMENTS

Compressed and pushed on every side, the weight
Crushes on flossing trusses, made from heat
And all the kindling gone. The hour is late –
But not yet struck. As Neon stocks deplete,

The star hiccoughs. Its throbbing skin pulsates.
Alfvénic waves roll out and in. It gleans
For food and – finding nothing – oscillates,
Then settles back. A huge contraction means

The end is close – but not here. Oxygen
It seizes now, another meagre fuel
Of little worth compared to Hydrogen
Of happy memory. This insipid gruel

It burns to Silicon and also Sulphur.
Small are the gains; gone, any hopes to stave
Away the drop which rushes to engulf her.
On Supernova Row, no plea can save

The star dragged on towards the final fate
The laws of physics have ordained and set.
The arc of time has passed; we now await
The culminating seconds – but not yet.

ORIGINS

While aged men and women fade to bones –
Their shrink-wrapped skin adorned in cotton flakes,
Their minds abandoned to that known unknown
Whose sole redeeming gift is that it takes –

An aged star remains a mighty thing
And blazes large and hot as death presents.
This flaming Phoenix roars – and dies to sing
The cosmic canticle of elements.

Chlorine, Potassium and Argon form,
Waste from combusted Oxygen – gas poison;
The butter metal; Nobel gas. The storm
Of fresh creation rages. Noise from, noise in

The deep reverberates throughout the star.
An end, a new beginning – still not yet.
All times to come recede absurdly far;
All times elapsed, lost dream-wisps to forget

As past and future sublimate to now.
It burns, it burns – and Oxygen is gone,
Gone from the core, dropped from the why and how
Of this last desperate struggle. Silicon

ELEMENTS

Replaces it, at once both last and least
Of all the fuels to feed this flailing giant.
At breaking stress, the yield again decreased,
The star is briefly, hopelessly reliant

On Silicon to hold it up. One day,
Or two or three, will see the glutton burn
And dissipate this all: the final stay
On what awaits. Its fagged-out ashes churn

The core to jumbled slag of Iron and Nickel.
Like some great torrent which once fed the lands,
Reduced to first a stream and then a trickle,
Fusion now slows, then stops. Skeletal hands

Point at the twelve: one last recap before
The knell on which the star will be destroyed.
A star needs heat to live. In search of more,
It fused the fused and, doing so, employed

A panoply of elements from small
To large. Each single merger liberates
Some energy and, with it, heat; and all
This heat supports the star while it creates

Still further elements. But these are bound
Deeper and tighter than their parents. When
They fuse, they give up less. Research has found
That counting up from lowly Hydrogen,

Ascending through the Periodic Table,
The larger nuclei are more confined.
They hold themselves in; they become more stable
Right up to Iron, where their internal bind

Attains its maximum. To either side –
Cobalt and Manganese – the grip is weakened.
The atoms – star-struck, trapped on fusion's ride –
Merge on and up; ascend this ferric peak and

Halt – briefly – at the summit. More would be
An endothermic step – and suck heat in,
Not give it out. The star's biography
Now comes to climax; its last rites begin

As Iron is forced into existence. Last,
And built to last, no mighty forge of Vulcan
But sulphur-bowelled Hades saw Iron cast,
Throes of a dying star. It piles in bulk and

ELEMENTS

All fusion stops. The source of heat runs out.
Above, the weights of time and famine press
Down on the burnt-up core. This, knocked about,
Squeezed by the crush of crunching, pinching stress,

Tries for a quantum hold. But Chandra's limit
Is far exceeded. Waiving all resistance
Against collapse, the core gives way. One minute
Now feels eternity, one solar distance

Like half the universe. The inner core
Shrinks in and down; the rest drops in its place.
Faster than any bird, it swoops before
It strikes and, plummeting, attains a pace

Almost one quarter of the speed of light.
Meanwhile the dead core, on its last descent,
Climbs to new peaks in density and might.
Electrons merge with protons; each event

Emits one hard neutrino – which escapes
To bring its image of impending doom
To far-off worlds, through brutal astroscapes
And real-time pictures of this closing tomb.

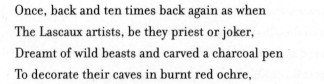

Once, back and ten times back again as when
The Lascaux artists, be they priest or joker,
Dreamt of wild beasts and carved a charcoal pen
To decorate their caves in burnt red ochre,

A star endured this fate. Its core collapsed
Off in the Magellanic Cloud. This made
Neutrini. They dispersed. As time elapsed,
There rolled through space their ghostly cavalcade,

And when I was six years old, it reached our Earth.
As fleet as fairies, most just zipped straight through,
But twenty left a mark — marks of great worth.
They showed that what we thought we knew was true:

That in this final phase, the stellar centre
Becomes a close and tight-packed neutron ball,
A nuclear tortoiseshell, perfect preventer
Of particles' attempts to breach this wall.

ELEMENTS

Shorn of support, the star's outlying parts
Drop under gravity towards the core.
Static at first, the nuclear coffin starts
To pick up speed. Accelerating more,

The star's whole history now cascades in.
Faster and then faster it concatenates
Both past and present in one deathly spin
Towards this core no force can penetrate —

Like cavalry who snort gunpowder air,
Canter in unison to the attack,
Hit at full gallop an unbroken square
And — not yet! still not yet! — are hurled straight back.

And now it blows — blows with tremendous force,
The force of ages, both those long since passed
And those to come. The lash lets rip the source
Of elements; the whip-crack outwards blast

Makes even more. Unleashed, the neutron flux
Pours on the nuclei exploding out
A flood of particles. Like rain which chucks
It down in buckets, drenching all about

ORIGINS

With skin-smeared drops, this neutron fire-hose
Scorches all. Hadron-drunk, the nuclei
Distort like mutant Calibans, as blows
Convulse them through. Collisions multiply

These awkward higgle-piggle neutron piles.
Misshapen, bloated and radioactive,
They cannot last. Though born extremophiles,
They soon collapse to hunt a more attractive

Configuration. Fission makes some split
And shatter. But decays that matter most
See neutrons turn to protons and emit
One lone electron, one chimeric ghost

Neutrino and…and nothing more. This switch
Sees total nuclear charge nudged up by one.
This alters element; and forms one which
Is less unstable — and what once is done

Is swiftly done again. With each decay
We move along a path whose final end
Must be a stable element. This may
Be Gold, or Tin or Tungsten — or extend

ELEMENTS

Up to Uranium. They all fled stars
Ungainly heaps of neutrons, whipped and battered
Far out of equilibrium. Their scars
Healed up; each found its element while scattered

In vacuum. An astronomer's cold eye
Observes. She looks. She censures with her gaze
This star, which nursed its brood of nuclei,
Just to expel them all and self-erase.

It was; then it was not. It leaves a shell
Expanding out from what has been, a vast
Trail of exhumed debris, a harrowed hell
Of freak-show daughter elements. The blast

Propels these through the void – and as they go,
They sow the dust with diamonds. Unimpeded,
This space tsunami-torrent spreads, a flow
Whose all-consuming force is not exceeded

In known events: not here, not anywhere.
But in its wake, life finds its cornerstones:
Boron for flora; Oxygen for air;
Iron for blood and Phosphorus for bones;

ORIGINS

Nitrogen, needed part of DNA,
Is fixed by plants out of the atmosphere.
And when we eat our veg, our five a day,
Our bodies take it in. Green stalks appear

When seeds are fed with falling drops of rain:
Di-Hydrogen monoxide. Icy pools
Freeze from the top: behaviour which sustains
Life's nurseries. Attached to molecules

Like whiskers, Hydrogen attracts — and holds
Together paper books and cotton clothes.
Lose it — the greatest stories ever told
Dissolve to dust while all decorum goes;

Then Calcium — required for film-star teeth,
Sparkling like vanities in theatre lights;
Plus, guards of belching bloody guts beneath
The skin, the killer ninja phagocytes;

Magnesium — whose bond to ATP
Creates a sort of micro-gasoline
Pumped through all cells to give them energy;
Vanadium for rodents; Iodine

ELEMENTS

For us – who break down fats with Chromium;
Bromine for corals, treasures of the seas.
The brain wants Zinc; nerves need their Sodium;
Fluorine for teeth; a pinch of Manganese

Clots bleeding cuts of arteries and veins,
While membrane gates, from single oocytes
To neuron networks sprawled through grown-up brains,
Hinge on Potassium electrolytes –

And as their counterbalance, Chlorine ions.
The next one's oddly named, yet I attest:
Only when baby lambs lie down with lions
And children build their dens in cobra nests

While tanks are turned to teapots – can we spare
Molybdenum. The B-12 vitamin
Is built around a Cobalt matrix; hair
Consists of Sulphur-laden keratin;

We need Selenium for healthy thyroids –
Plus Carbon, Carbon everywhere, the base
Of earthen life from elephants to viroids –
Till HAL 9000, Adam of its race.

ORIGINS

And not just these, but rarer elements
Key to the furniture of modern lives.
The nibs and inkwells, old accoutrements
Of social intercourse, now gone, life thrives

In message groups and rightward swipes. Your phone
Preens out its peacock colours on the screen;
Vibrates itself; trills out a dialling tone;
Looks up the freezing point of glycerine.

Its magic witches' tricks require within:
Europium; Tungsten, Ytterbium;
Silver and Cadmium; Bismuth and Tin;
Tantalum, Mercury; Lead and Erbium.

What did it take to find and separate
These oddball jewels out of trumpery
Then build the circuitry to aggregate
Them in this apex of technology?

One: hours of darkness. Whiffs of week-old sweat.
Cold stacks of jagged rubble, piled to soak
In acid baths. Strange odours to forget
Of pungent chemicals and acrid smoke.

ELEMENTS

Think of those factories and smelting plants
Which tease these wanted metals out of ore
Hauled up from underground by human ants,
Who pay with schooling, hygiene, health – and more.

Two: hours of darkness. Whiffs of week-old sweat.
Cold stacks of pizza greet the breaking light
Of dozy-fingered dawn. The man-drones get
Another coffee, then sit back to write

The algorithms and long lines of code
Which comprehend the whims of humankind
And translate 'Peotry: less ravelled rode'
To bring up Robert Frost. Call all to mind,

Think of the toil and sacrifice which goes
To set all knowledge free. With what results?
The viral spread of online videos
Of cats on skateboards and unclothed adults.

But great heroic tales, from Fionn to Thor,
Feature with prominence those elements
We do not need, and thus we value more:
They make and prettify the ornaments

ORIGINS

Which give our lives a life beyond the stress
Of animal survival — and, of old,
One was the most cherished for loveliness.
It lasts. It never fades. Its name is Gold.

When Solomon received the Queen of Sheba,
His banners flew, his trumpets blared. His own
Servants salaamed, styled her 'Your Grace Sahiba',
But he ensured she saw his golden throne.

When proud King Midas was allowed one wish,
All lesser metals left his feelings cold.
He wanted every bowl and serving dish
To gleam of riches, and so gleam in gold.

When tasked to visualise all Heaven's host
With haloes hallowing long lines of saints,
Artists picked out the shade that fitted most
And drew their holiness in golden paints.

And when, in idle mood, I touch and twiddle
Round and around again my wedding ring,
The living flesh thrust through its pensive middle,
I see, reflected from this little thing,

ELEMENTS

The buried quarks inside, the Hot Big Bang,
The stars of cosmic dawn which woke the sky,
The death and life, the stellar yin and yang
Of supernovae birthing nuclei.

And in a bustling common-room I meet:
Coulomb and Paracelsus, locked in talk;
The chirpy Hoyle, who runs with haste to greet
Newton in tears of joy, eyes on the chalk

In Chandra's hands; Sir Arthur cheers him on
To fill the blackboards; the attentive stare
Of Fowler follows. Next, I come upon
The Burbidges, in convo by the chair

Where Einstein naps. Cecilia Payne, a sea
Of smoke, prepares to give a catch-up class,
Her pupils Ibn Rushd and Ptolemy.
The hallways fizz with learning. Here I pass

Saha and Fermi, Feynman, Weinberg sunk
In new discoveries and calculations;
Brahe shocked sober; al-Khwarizmi drunk
On fruits of centuries of observations.

ORIGINS

Assorted Nobel Prizes line the wall,
While on the balcony the telescopes –
From Galileo's to James Webb – recall
Those times when knowledge overwhelmed all hopes.

Here dwell my colleagues; here I find my tribe,
Here boundaries of age and nation end.
Grant me one foolish boast, one shout of pride:
I know them all since youth and call them friends.

This vision fades; the daffodils return,
And 4x4s snug at the Barley Mow,
Near Thames-side cottages for those who earn
Enough. But look, the stars! Be rich; and *know*.

Galaxies

Preface to *Galaxies*

This poem describes the origin of the galaxies and other cosmic structures that fill the universe — in all directions and as far as we can see. In our modern understanding (dating from the early 1980s and with progressively more support since then), such structure in the universe arises from quantum-mechanical fluctuations on tiny scales in the very early universe, subsequently stretched to enormous distances through a period of *cosmic inflation*. During this inflationary epoch — which ended long before even one pico-nano-microsecond had elapsed — the universe, governed by the laws of Einstein's theory of general relativity, grew enormously in size. Originally only present at tiny lengths, the quantum fluctuations were stretched in size to become the seeds of future galaxies.

The poem describes this and splits into four main parts. The first part introduces the cosmological puzzle of why the universe is so alike in all directions. The second part introduces the idea of cosmic inflation and explains why it ensures the homogeneity of the universe on the largest scales. The third part describes one of the most appealing aspects of inflationary theory: the way that it not just ensures the large-scale homogeneity of the universe but also contains a mechanism to generate, through the uncertainty principle of quantum mechanics, local structure and inhomogeneities such as galaxies. This contains a verse biography of Werner Heisenberg, one of the founders of quantum mechanics who discovered the uncertainty principle. The fourth part, a coda to the main poem, is an extended simile which relates inflation to the way social media can amplify temporary ephemera and give them permanence.

Galaxies

Like sexed-up wildebeest, youth congregates
From Eros down through neon streets, in fear
Of nothing worse than age. Initiates
Of adulthood, they feel its vibe; they hear

Its call – of pubs and clubs, of bars and shows
Which burst with darling lovelies everywhere.
Their nightly hormone-fuelled migration flows
Through private drinking holes and public squares.

The rutting throngs dispatch their roving eyes:
In front – through the electric ersatz haze;
Around – to scout attractive girls and guys;
Down – throats of narrow dive-filled alleyways

Where lurking hoodies flog amphetamines;
Not up – here where the city never sleeps,
The light which slew the shadow gloam-goblins
Has also slain the stars. Just language keeps

ORIGINS

That past alive, a past where dropping dusks
Announced each coming night of who-eats-who
And God-knows-*what-was-that?* Through words, the husks
Of our ancestral fears survive. They knew

Terror – the sprites and ghouls outside, who swarmed
To seize and sacrifice their children. Dread
Kept them in caves, near smouldering fires which formed
Their shield against the dark. But, overhead,

They also knew eternity, the sheen
Of that extravagant divine display:
Sombre yet bright, majestic though serene,
Coruscant but unchanged, the Milky Way.

For who then sensed what vistas would be gone
Or understood the wealths we were forsaking
When Thomas Edison and Joseph Swan
Gave us their gift of light that keeps on taking?

Back to the start: the oldest known profession
In science has the name 'astronomer',
A guild whose work reveals the deep expression
Of strong desire – to quit the plough, prefer

GALAXIES

Hard fact to old wives' tales, then spend eternal
Rapt nights admiring Heaven's bodies, free
Of flesh-borne wants. They first observed diurnal
Rhythms in sun and moon; cyclicity

Each lunar month; the slow and gradual movement
Of planets past the stars. As naked eyes
Reshaped to coated glass, they found improvement
In clarity, detail and depth. The skies

Of Ptolemy now changed; with Earth dethroned
And Sun debased, while Jupiter grew moons –
And all who, drunk on vintage books, disowned
This new cosmology were dumb buffoons

('Like you, your Holiness!' quipped Galileo:
And ever since, there rose not only glory,
But grumbles also, *in excelsis Deo*).
Although confined to house arrest, the door he

Had found now filled with travellers. They gave
Their hearts to space; their love to lenses. Proud
To grind and proud to polish, when concave
And convex worked as one, skies stripped of clouds

ORIGINS

Saw them transpierce the midnight veil. These squints
Revealed entire new galaxies, out far
Beyond the Milky Way. Such smudges, hints
Of *something*, spurred them on. At Palomar,

At Mauna Kea and at Siding Spring,
They hauled their telescopes up mountain paths,
In hope that eerie solitude might bring
Faint photons safely in. With spectrographs

And twelve-inch plates, with coffee-haloed lists
Of all atomic lines, they came to know
The stuff of which our universe consists,
What form it takes and where. One faint red glow –

A quasar flickering – told them of deeper
Discord, through violent pancatastrophon
Perceived from far away. Is there a steeper
Vertigo than to sit and think upon

Such intervals and what they mean? To kiss
The waiting retina, its photons spent
One billion years en route – yet during this
Abyss of solitude, each week that went

GALAXIES

They passed one hundred billion miles. Somehow,
They reached our Earth; received, announced in state
As time's ambassadors from then to now,
They told of agonies which long predate

Tyrannosaurus. From one quasar sprung
A host ten thousand strong, at monstrous distances
Which raze our senses; all were placed among
That class of objects whose entire existence is

A mystery. Decades of observation
Were turned to catalogues of where and what,
With which the curious of every nation
Considered what was known and what was not.

Reflecting on those darkling years, immersed
In nightly peep-and-stare, such seekers reckoned
That two great questions were in play. The first,
How could the cosmos be so large? The second,

How come it looks so much the same? For Northern
Astronomers, eyes tight to telescopes,
Saw space weep pearls of galaxies; far more than
They could enumerate. Those isotopes

They saw, they recognised from Earth. Their stars
Could slot inside our Milky Way. The sight
Contained familiar spectra: magnetars,
Plus cold white dwarfs and hot red giants. This light

Had travelled for ten billion years, so great
A journey to encounter childhood friends.
Why does our local cosmos replicate
The physics of its furthest Northern ends?

Meanwhile, in Southern skies beneath the tropics,
Astronomers, eyes tight to telescopes,
Saw space weep pearls of galaxies, the optics
Beyond all numbering. Those isotopes

They saw, they recognised from Earth. Their stars
Could slot inside our Milky Way. The sight
Contained familiar spectra: magnetars,
Plus cold white dwarfs and hot red giants. This light

Had travelled for ten billion years, so great
A journey to encounter childhood friends.
Why does our local cosmos replicate
The physics of its furthest Southern ends?

GALAXIES

To North and South, to East and West, as far
Back in the past as eyes can see, the same
Objects are found – a fact which seems bizarre
When we reflect on it. These twins proclaim

Some common origin, some long-gone past
When all were joined as one. But when we trace
Their histories back in time, out through the vast
Expanse of space, they never meet. The place

Of union does not arrive; the locus
Which birthed them all is never found. Instead,
They miss. Reverse Big Bangs may seem to focus
All to one point: may *seem* – we are misled

Without the maths. The singularity
Awaits; but until then, all stay apart
And do not meet. The spacetime geometry
Can be rewritten – artifice and art

Conspire to write down new coordinates.
These, morphing spacetime, show what causes what
With utter clarity. The *ifs* and *buts*
Of slushy feeling dissipate, their knot

Of *oughts* and *shoulds* dissolved in algebra's
Therefore and *will*. No distant galaxy
Has ever met its close comparators.
That is all; question finished; Q. E. D.

Except – behind every successful proof
A key assumption skulks, the cornerstone
On which the argument was built. No truth
Rests as an island left to stand alone,

Apart from other knowledge. In this case,
The premises include the supposition
That space, through all its backwards breakneck race
Towards this mad dream of a theoretician,

The cosmic singularity, contained
The matter that we know and only that.
Wrong: this key claim is false. The unexplained
Coincidence in looks, the samizdat

Galactic copies spread like litter – all
Arose from early rapid acceleration
Which spun our universe from minuscule
To massive, quick as that. It's called 'inflation'.

And what is this? It is the answer to
Our riddle pair: of how the universe
Can be so large yet so alike. If you
Approve of theories grandiose but terse,

Which draw on maths but found themselves on data,
Which come from now but focus on the past,
Which set themselves to rival the Creator
But let a numinous transcendence last,

Then hear my simple cosmic story of
A time long, long ago. This narrative,
Whose later chapters credit Mukhanov,
Steinhardt and Linde, first appeared (though with

Some parts in earlier work by Starobinsky)
In 1981 from Alan Guth.
Its verified predictions have convinced me,
And most cosmologists, that here is truth.

ORIGINS

This was a time of majesty, a time
When vacuum – empty space – was tightly packed
With energy. For in this paradigm,
Although the infant cosmos may have lacked

Electrons, photons and all normal matter,
It had itself. And in itself, it found
The strength, at first to flex and then to shatter
Those geometric chains which shackled round

Its reach and held it small. Not from outside,
Not through external help or human aid,
But from its own reserves, it amplified
Itself. As space divided, more was made

To fill the gaps: like sourdough starter yeast,
Whose spoonfuls have infused five thousand loaves
And made each rise in zest, but has not ceased
To be. Its youth regenerates; it grows

And bubbles out. Convulsed with potency,
Ert and dynamical: such are the laws
Of space in general relativity.
They bear the marks of Albert Einstein's claws.

GALAXIES

There lived a man of old, who strangled lions
With his two hands. He seized foxes, to tie
And fire their tails, then slew great hosts of Zion's
Enemies. Captured, bound, gouged in each eye,

He sacrificed himself to thwart his foes.
His name was Samson; in his hair lay strength.
Today, our sense of genius also owes
Something to unkempt hair of untamed length:

For, asked to visualise appearances
Of those who excavate the depths of space
And mine its logics for analysis,
We picture Albert Einstein's smiling face.

With Europe's youth minced to cochonnerie
To feed the gaping mud maw of Verdun,
He sought to find the laws of gravity
Through simple thought experiments: someone

Is falling from a roof. If she is blind
And cannot see the ground before she hits,
Can she detect, with her still-active mind,
Some trace of gravity? He racked his wits

And answered, 'No.' Or take a man inside
An elevator, lifted (God knows how)
To space, with towing cables firmly tied
To it. A spaceship takes the strain, to plough

The empty vacuum. In his can, can he
Perceive the slightest form of deviation
Between a downward force from gravity
And driven uniform acceleration?

Never – never – never! They are the same.
This total absolute equivalence,
Promoted to a principle, became
His key to gravity. With elegance

Of mathematics, with economy
Of physical law, he reframed the forces
Of Newton, turning them to geometry.
So take the field around a star – this source is

Reincarnated and reborn a dimple
Upon the skin of spacetime. To his eyes,
The science shone succinct; its laws were simple;
For Riemann's tensor calculus supplies

GALAXIES

The needed algebraic formalism
For physics on curved spacetime manifolds.
Gravity, viewed through this conceptual prism,
At first seems blurred. But then, the image moulds

Itself and sheds the fuzz. Ideas reshape
And come to focus. Forces, rearranged,
Morph to a geometric mathoscape
In which both all remains and all is changed.

In geometry and space and time, he saw
One entity, whose laws must both obey
And govern gravity, one setting for
First action then response. But what were they?

As Michelangelo, sat on his stool,
His chisel seeming merely fit to scar
And notch and peck, had with this humble tool
Transformed dead marble to the *Pietà*,

So Einstein, at his desk, sat on his chair,
Took up his writing pen and artfully
Sculpted the empty sheets of paper there
To ten equations: the equality

Of $G_{\mu\nu}$ and $T_{\mu\nu}$. Exact,
And perfect; proof that God's a sybarite
In mathematics. Take this single fact,
And read from right to left, or left to right,

For matter showing spacetime how to curve,
To mark its pressure and its energy,
Or spacetime telling matter where to swerve –
Although, in Einstein's form of gravity,

What's bent is mutable. Our Earth orbits
The Sun elliptically – so textbooks state.
But Einstein took and flipped Newton's ellipse –
In four dimensions it runs ramrod straight.

Now *this* was genius; to reinterpret
The shoves and pulls of bare mechanic force
Hard buffeting the planets round their circuit
As unimpeded flow on constant course.

Both Newton's apple and Apollo's sun
Respond to gravity as of one mind.
Their routes through curved spacetime serenely run
Along the shortest straightest paths they find.

GALAXIES

So astronauts who orbit near the Moon
In monthly passage round the Earth, in turn
Bound to its liege the Sun, a picayune
And common ball of gas whose path, we learn,

Will circumnavigate the Milky Way
Once per two hundred million years, spun far
Out through its spiral arms locked in the sway
Of Lord High Sagittarius A*,

Perhaps reflect on their trajectory
And how, although their route appears to curl
With quite exuberant vorticity
And epicycles piled on cycles – whirls

Thrown like a sparkler's fleeting dying dart –
To keener eyes, their road instead defines,
The swirls transmogrified by Einstein's art,
The most direct and straightest of straight lines.

On top of this surprise, one final twist,
Key to inflation and its physics, lies
In Einstein's theory: nothingness consists
Of something, doctrine I shall summarise

As 'Not just matter matters'. Empty space
Need not be desolate, but may abound
With latent vehemence whose surface trace
Is vacuum energy — whose heft redounds

Back onto space to form a feedback loop:
What one enacts, the other back-reacts.
Locked in this physics bitch-fight, in one swoop
They rouse the dormant void and supermax

All its dimensions. Pardon? Space can make
And replicate itself? This power seems
Bizarre — so Einstein said, 'My worst mistake!
I shall forbid it.' But these odd regimes

Indeed occurred when time was young, the first
Epoch we can believe in, when a grain
Of space became a bushel's worth, dispersed
To seed the ever-growing cosmic planes.

What governed this? What caused its fertile vigour?
From where was found the lusty driving strength
To cut to spacetime's heart, then drag it bigger
In next to zero time? What growth! The tenth

GALAXIES

Power of ten does not come close. The whole
World that we know, our nourishment since birth,
Can offer no comparison. Our sole
Analogies, our similes of Earth,

Are feeble mimics. Seeds – with lazy ease
They danced for miles disdaining walls and gates,
Jigging out raucous reels to surf the breeze,
Then grew to gnarled windbreak agglomerates.

Or, reader, think of us: we were the size
Of single embryos which must be viewed
Through microscopic artificial eyes –
Cells which became ourselves. The cosmic mood

Discards with scorn such common stuff. For space –
Doubled, redoubled, eighty doubles more –
Full disembowelled the very sense of place
As future overwhelmed what came before.

What does this mean? Can we envision this
And frame such grandeur on domestic scales?
Or is it like expounding calculus
Before an audience of drowsy snails?

ORIGINS

When folding paper many times, one sheet
To book counts eight – and stop. Try if you must –
A further twelve would pass a hundred feet,
An extra thirty reach the lunar dust.

From bedtime tales, recall the shah who gave
One grain of rice today, with two tomorrow,
Then four, then eight, for sixty days. A grave
Mistake: the nabob king learned, to his sorrow,

This promise had committed all his wealth
And more besides, enough to drown his land
Five metres deep in rice. This tale itself
Should shock, but tenfold once we understand

How much it undershoots what nature did,
How faster – higher – stronger – was the push
Of spacetime out and through itself. Its lid
Was blown, then blown again, the cosmic whoosh

Depositing those infant galaxies –
Like cuddle-babes asleep in Lyonesse,
Snuggled in cosy new-born nurseries –
Out to the furthest ends of nothingness.

Though called 'inflationary', it's not the kind
Loved by the greybeards of Threadneedle Street.
No quarter-points to thrill their pin-striped minds,
No baited breaths as central bankers meet.

What physics dares and does is as Versailles
To Bauhaus subjects hunched by their reliance
On human labouring; one reason why
We call economics a bysmal science.

For it would treat the turbulence of Weimar
Like Götterdämmerung threshing the state,
Which, with contempt, we shall dismiss as minor
Twiddles around a zero-interest rate.

Growth such as ours would turn a penny stock
To Google riches worth a googol-pound
Before one tick of an atomic clock
Could tickle ears which hear atomic sound.

Abundant glitzy wealth is comeliest
Spent for the public good, a lavish hand
Dispensing blessings from above. The best
Largesse of all is understanding – and

ORIGINS

This growth permits the single ancestry
That we have sought for galaxies. Together,
They start; apart, they end. Like sesame
Seeds grabbed by wind, they are dispersed — and never

Again are found in such proximity.
But once, in the beginning, all were close
And huddled, when each proto-galaxy
Shared one environment. Not grandiose

As yet in looks, frail wisps of slightly more
And slightly less, here they imbibed the same
Communal atmosphere. The same: and for
This reason, when such embryos became

Galactic giga-stars — the homes between
The dead expanse of void, the welcome lamps
Among the solitude — their grizzled miens
Retain to now that common early stamp.

So, whether curious in child-like thought
Or seeking that elusive parent-pleaser,
We learn that time and scholarship have wrought
The resolution to our teacher-teaser,

GALAXIES

'How can all stretch so far across the skies
Yet stay so similar in visual style?'
The early universe, we find, supplies
The elements required to reconcile

These odd and clashing facts. Cosmic inflation
Connects the two; its logic chains provide
A full, yet parsimonious, explanation
In which no entities are multiplied

Beyond necessity: the Occamist
Exults with pride, and so do we. And yet,
As egg and sperm precede the blastocyst,
As oral speech precedes the alphabet,

These starting micro-overdensities
Are not the start. They have an origin
And came from somewhere — but their geneses
Lie far from Einstein's gravity, within

The subatomic quantum world. So drop
The grand; throw down the great. We turn around
And start an odyssey which only stops
When snug inside the atom, where is found

ORIGINS

The physics that we seek. The universe,
A hundred billion light years in its size,
We leave. Those clusterings which intersperse
Its voids, great groups of galaxies, comprise

Simply the first of many staging posts
Which punctuate our inward journey. Gas
Marks these; a half-of-nothing wispy host
Whose volume gives it heft; their total mass

A million billion suns. Zoom in and down:
Each single galaxy a habitat,
Both stellar capital and shanty town,
And home, perhaps, to alien life (but that

Is not for now). From galaxy to stars
And solar systems – from this crowd, pick one
Of many look-alikes. All similar,
But one is special: round it, planets run,

Gas giants further out, rocks closer in;
The number three enchants with blues and greens,
Eight thousand miles across, a mottled skin
Of mixed-up soil and water, battle scenes

GALAXIES

Of eaters and the ate – and here we think
First of ourselves, of *Homo sapiens*,
Five feet or six, or thereabouts; then shrink
Our scope one thousandfold, as through a lens

We stare at louses' gobs and beetles' thighs
(O Entomologists – I envy you).
We blush and squeeze inside; attention flies
To cells and their machinery. Our view

Takes in these bio-factories; their gates
Now barred, now opened up as messages
Zip to and fro. The centre regulates
All and commands: Look here! This protein is

To be constructed. You! This DNA
Needs copying – transcribe it well, then draw
Up plans to split in two. Swept down away,
We zoom to smaller scales as more and more

Each cell becomes a giant. We reach the realm
Of viruses, that Moloch's den of thieves
And genocides. Perhaps these underwhelm
In their physique; the slender build deceives

ORIGINS

As to their wickedness. Here Smallpox lurks,
To taunt unchilded mothers with the tears
Of pus-filled death, as brother Measles smirks
His broad infectious grin. Their double leers

Rejoice in misery and grief, in what
Was meant to be but never was. They greet
Their sister Influenza. She wears not
One stitch – except an old used winding sheet

Wrapped and rewrapped in fifty million folds.
Here dwells Rubella, scourge of growing babies,
And Chickenpox. Too cool for common Colds,
Ebola cackles next to his mate Rabies,

Who writhes with laughter as he speaks about
His screaming victims; how he found their minds
And drove them raving mad from inside out.
We leave, rush down in search of peace and find

The quantum world. This lives on nano-scales,
Inside the atom, far beyond the sight
Of life, which can ignore the fairy tales,
The mind-befuddling scary tales, of light

GALAXIES

At once a wave and particle, which sees
Itself and interferes; both there and here
And here and there, it ghosts through walls with ease
By tunnelling. Here flourish wild ideas,

In which the solid stuff of common sense
Dissolves to quantum fuzziness. A smudge
Is now a particle: thus we dispense
With everything we thought we knew. The judge

Of what exists for real is now no more
Our finger-feels; the squeezing comfort-touch
Which reassures us children that the floor
Of our perception is not mulch. But such

Conceits of knowledge we strip off. We lose
Our native intuitions, as we free
Our minds from what is obvious. We choose
The scars of truth. We read, and learn to see

Electrons smeared and spread like cottage cheese
Yet holding energy in discrete lumps.
Packaged in orbitals, their calories
Are given up and gained in quantum jumps.

ORIGINS

We scrap through hostile tomes of symbols writ
In ancient alphabets; but ϕ and ψ
Now serve new masters. In this modern script
They mark those wavefunctions which signify

All that we know, all that we have, the sum
Of all we might perceive. As mental blocks
Appear, we reason past them – and so come
Upon a great apparent paradox,

Whose subtleties surpass mere calculation.
The path is tortuous, the route unclear.
Though central to the physics of inflation,
We halt before this maddening idea

Whose mastery seems unaffordable
When priced in thunken thoughts (a 'neuro-erg'?):
Quantum uncertainty – the Principle
Of Werner not-a-Nazi Heisenberg.

GALAXIES

His start in life the gods both blessed and cursed:
Born a Bavarian, 1901.
He loved his physics and this love was nursed
By Sommerfeld at LMU. A son

Of Germany, he liked to hike and walk,
Enjoyed outdoor pursuits, roamed with his scouts
Through winding mountain trails, where he would talk
Of physics, Plato and the dreams and doubts

Which haunt young minds that fizzapop with sparks.
In 1919 he joined the Freikorps
To fight the Anarchists for teenage larks.
But then in '22, he met Niels Bohr,

The Danish guru of the quantum world.
The boy became a man; from Bohr he learned
About the new and strange results which curled
Up out from atoms like white smoke. He turned

To them with focused zest. These teases hinted
That laws which undermined all Newton's physics
Lay barely out of sight; but when he squinted
To read the detail, the desired specifics

ORIGINS

Were merely blurs – until he took the boat
To Helgoland, to gain a brief respite
From hay fever. The air which soothed his throat
Unblocked his mind, and late one summer's night

The answers came – in a joyful tumbling rush,
Like undammed streams which chortle past the rock
That had confined them. Through this moonlit gush
Of sparkling clarity, come three o'clock

He had the problem solved. He met the dawn,
Which gleamed with virgin light and understanding,
And set out to explore. That night had torn
The veil of ignorance, the rip expanding

Up out from atoms to the principles
Of what a measurement can even mean.
From such a start, there followed thesisfuls
Of exploration. Heisenberg had seen

How nature scanned, had found the basic rules:
One single form – but rich enough to spawn
All modern physics, through the varied tools
Of Pauli, Schrödinger, Dirac and Born.

GALAXIES

Such youthful innocence! What bliss to live
From 1925 to '27!
As ancient certainties fell through their sieve,
These quantum younkers felt themselves in heaven.

But down on earth, a hellish broth was brewing
In febrile post-war Munich's beer-swilled cellars,
Where hearts of Rat with weasel tales were stewing,
Stirred by brown-shirted fork-tongued storytellers.

'National disgrace! See how the Heimat bleeds —
Such endless tears! Well, think! Whose thirst for money
Drove us to this? The Jews! Those sly half-breeds
Pinched all our Land's abundant milk and honey.

Now look around: I see you army men —
Strong in the field; with grit, with pride you fought —
Dumped and betrayed for gold. That vipers' den!
Profit and purse-strings fill their every thought.

No more! I will reverse our nation's fate.
Support with arms my struggle. Come! Fulfil
Destiny's call — and make our Land a great
Power again. One race! One Reich! One will!'

Nothing was safe. No part of public life
Survived unscathed. This broth, swallowed whole, fed
Deep-buried long-nursed hate and belched a strife
Polluting all it touched. This swelled. It spread

Like cancer, its obscene far-reaching growth
Invading outward from the starting site.
Even professors took the Führer-oath,
Stood for 'Die Fahne hoch' and vowed to fight.

No fool, no Nazi, clever Heisenberg
Shunned guff like 'racial science' and 'Deutsche Physik'.
Despite such frauds – and torch-lit Nuremberg –
Patrial love defined his sense of civics.

So at the crux of where to use his brain,
He saw himself a German through and through.
As darkness dropped with knives, he would remain
But surely would have fled if born a Jew.

But he – no buts! His lasting reputation
Yields to his labours on a Nazi bomb.
When forced by fate to choose, he chose his nation
As Zyklon pyres were built to dwarf the Somme.

GALAXIES

Reader! – how tempting, after eighty years,
To haul the Book of Judgement from the shelves,
Reject internal angsts and human fears,
And damn those choices we were spared ourselves –

Such it has ever been; such will it be.
His culpability be what it may,
So long as minds still think and eyes still see,
His Principle will stand. What does it say?

A particle – a fermion or boson –
Is measured, to pin down two facts which ought,
By sense, be knowable: how fast it goes and
Where it is. Dream, invent; deploy all thought

In vain: to know them both demands a change
Not of technology but nature's laws
As, though this seems unfair and monstrous strange,
This is impossible. Its strictures pause,

Now and forever, routes to scale the top
Of man's ambition, the compulsive greed
To know it all. Instead, they make us swap
What we may know of place with that of speed.

These two are tangled; we can pick just one,
As with a lover's diary and their trust.
Arrange it how you like, but when all's done,
Speed or position blurs: for those who must

Seize one for certain surely lose the other.
Each can be had alone, but not together,
And doubting smarty-pants who try discover
That Nature shakes her head, and spits back 'Never!'

So says his Principle: what does it mean?
In quantumland, there is no place of rest
To sit in stillness, no idyllic scene
Of perfect stupor, no oasis blessed

With idle laziness. Always there's motion,
Always there's energy, a twitching kernel
Of somethingness in spark above an ocean
Of void. A quantum bell-rope swings eternal;

A quantum motor never stops. The coldest
Vacuum, though drained of outside heat, still burns
With inner vehemence; all while the oldest,
And emptiest, expanse in spacetime churns

GALAXIES

With hidden bacchanalian quantum fury
Which is, shall ever be and ever was.
Not learned fancy, not the stolid jury
Of common sense may make it else – because,

Down in the dark-delved depths before all dawns,
These laws were weaved into the woof and warp
Of nature's fabric, knit with wiry thorns
Not silken platitudes. Such truths are sharp,

And razor-edged to scour soft minds. They root
Themselves in nature's firmament and rule
Both particles and space, first-budded fruit
Of time and universal stage. They spool

The many threads of all that is, in loops
Of meaningless meaning drawn straight by chance.
But why do galaxies and angels stoop
To care about the subatomic dance?

The quantum world is small – but not for long
Inside a universe expanding fast.
As first was last, as weak evolved to strong,
So minuscule contusions from the past

Grow to the sprawling cosmic skeleton
Around which structure forms. Behold today's
Clusters of galaxies; reflect upon
Their dazzling size; admire the hot X-rays

Which flaunt their beauty. But, beneath the glow,
Unpeel the many skins of time and pry
Into their origins. My God! These show
Quantum mechanics naked on the sky.

To watch this happening, zoom down to space
On nano-; pico-; zepto-scales. We slow
Down time, restrain its steps to crawling pace
And watch the hand which held the cosmos throw

The quantum dice to set how nature starts.
The playing pieces – particles and fields –
Are shaken up. The lot of chance imparts
To each the starting strength their number yields.

Thrown, and again; thrown, and once more. As thick
As humid air, as countable as thorns
In bramble-smothered woods, with every tick
Of every clock at every point there spawns

GALAXIES

Another quantum fluctuation. Tossed
With force, on tiny scales this dice face wields
Near absolute dominion – powers lost
In space's endless voids, where smallness shields

These secrets from the wider world. Within
The quantum realm, they flash, flare and excite,
Ignored and overlooked – till lengths begin
To stretch. The universe inflates. What might

Have once been close now separates, is dragged
Apart before it can return, is ripped
From inside out as space expands, is snagged
Upon an ever-stretching ladder. Gripped

By spacetime's vice, the small is forced to grow.
All childish ties of neighbourhood are torn,
Bound to this geometric path which flows
Up, up, above and out. And, thus, was born

Our quantum universe. Nothing can flee
The great expansion. Every fluctuation
Is swollen huge then etched indelibly
Across the cosmos. In this way, inflation

ORIGINS

Picks up the trace of quantumland, whose laws
Govern the smallest scales, and stretches it
On spacetime's rack; to set the primal cause
Of all the magic night-time skies emit.

But why? And whence? What physics underlies
Inflation's origin? We do not know
What came before, nor what strange well supplies
The energy to drive expansion. Though,

Can migrant swifts define magnetic force?
Can plants say what is meant by chlorophyll?
Do orcas want a compass for their course?
Do lions ask a butcher how to kill?

The dancer's feet will land where rhythm leads,
Faster than thought his trebles, stamps and taps;
Instinct provides the steps the music needs
And muscle reflex leaps past reason's gaps.

Spacetime responds. It does not ask permission
To dance the Linde-Guth. We calculate
As we are deaf, and only grasp position
In terms of algebra. We integrate

GALAXIES

From then to now; we search for formulae
Which turn initial states to final ones.
We look for tricks; we hunt simplicity
Through words to name and tame this crack which runs

Across, then splits the stage. In metric modes
And perturbations, in their grand advance
Through cosmic time, in mazy Python codes
To slide them onwards, we translate the dance

To runic scripts we think we understand.
These scratchings tell us that within the span
Required by light to cross one grain of sand,
This whirling jiggerbug of catch-who-can

Bequeathed to empty space three score exponents
Of growth, plus fodder for ten billion years.
We know the circuitry, but no components.
We know the governor, but not the gears.

So much! – but still, we want those plans, to trace
The inner rationale, sensed but not seen,
That drives the stark and adamantine grace
Of this eternal cosmic growth machine:

ORIGINS

Where raw ingredient of quantum vacuum
Is taken in at tiny lengths. This fizzes
With correlation, in the feisty va-voom
Of nothingness. The cosmic ratchet whizzes

These up and out to ever-larger scales
Outside the quantum realm, to freeze them in
As they turn classical. Each wave entails
One more, in constant growth whose origin

Is empty space plus vacuum energy.
Thus blooms our cosmic home, whose perturbations
Provide the modern macro-legacy
Of ancient micro-quantum fluctuations.

Profound, and true, ideas; but not so much
That only sober solemn words can fit.
Oppressive sermons and equations touch
The hearts of blessed few; but souls admit

Free through unguarded doors the zest of laughter;
The yarn, the fairy tale, the telling jest;
The evening songs before the morning after.
Souls take, and learn. And so, before I rest,

To those whose technicoloured minds create;
To those who dream; to those whose verses sowed a
Full field of seeds whose fruits cross-pollinate
All thought: to you, I dedicate this coda.

It takes our escapade along the route
From next to nothing into galaxy.
A bit of pixie dust; see it transmute
To a light-hearted social allegory.

Coda

As galaxies were born in quantum flickers,
Which started tiddly then were stretched ginormous,
This epic simile of knights and knickers
May, through its whimsy, tickle and inform us:

Our cosmic parable begins its telling
At lunch of rucola and mozzarella
Inside a genteel Georgian townhouse dwelling,
Home to Sir Robert and his Arabella.

ORIGINS

Here feng shui scents of pine and jasmine season
The lives of Princess Charming and her Mr,
Who – she suspects, and not without some reason –
Is apt to cast his eyes towards her sister.

A wince of sudden pain, an acid spasm
At summer barbecues and family Christmas
Greets every glance across that moral chasm
Where scent and evening light construct an isthmus.

To pass through thin white shirts and cotton dresses,
To reach down nippled bras and up suspenders,
To trace out panty lines with soft caresses,
To leave his flowerbed and gently tend hers,

Stretch thoughts not hands – for, though his pupils wander
In largely unconcealed appreciation
Of one both slimmer, bustier and blonder,
He holds his heart in tight. He sees damnation

And chooses faithful love, each year expanding
In breadth and magnitude (as with his stomach
And self-esteem). Saint-like in understanding,
She overlooks the straying eyes and hummock

GALAXIES

Most of the time. But even holy patience
Knows nights freighted with dreams of pretty faces
And days weighted under rotund complacence.
These worries nibble at those social graces

Which lubricate their marital relations.
But, though no more may his aura bewitch her,
Their union survives on firm foundations
Of twenty years wed for poorer or richer

And mutual charity, kind and protective.
But when they fight – as spouses will – their quarrels
Salt all the air with lashings of invective
And slurs dishonouring their minds and morals.

We join our couple as she quietly wishes
That after food she cooked, despite his millions,
He rouse himself to clear the bloody dishes,
Not just expound on Pound and Carlos Williams,

Such as he does today. Some guests are present,
Among them Sasha, an aspiring poet.
Our host beholds the room, admires his pleasant
Audience, sinks his second glass – and so it

ORIGINS

Begins. 'Our world awakes from hackneyed slumber.
Down, dead form; down, old childish modes of diction –
Like autumn slush, like crated rotten lumber
By steel and concrete towers. Scorn the fiction

That is tradition. Fire the soul; embolden
Your national Muse with fervent zeal and speed her
To destiny. Dare big: the fates unfold in
Our time and common people seek a leader.

Young writers, train your style through careful timing
Of line-breaks. But – for neither love nor money
Should modern poetry appear that's rhyming,
Rhythmic or readable (and *never* funny).

Instead, construct a text adroitly drafted
To be obscure and difficult: so readers,
Thrashing for meaning out of work thus crafted
(Like eagles pecking crumbs from sparrow-feeders),

Compelled to mine their brains and quarry deeper
Beneath all surface sense, can learn the flavour
Of unbreathed airs and find your voice the reaper
Of their naivety. Do me a favour –

GALAXIES

Teach them thought.' As he wipes his mouth and pauses,
She swoops between the napkin and the pesto.
'I'm sorry, Robert; but these rambling clauses
Reveal the problems with your manifesto.

It's not just *what* you say, but *how* you say it.
Pure seamless bolts of sound and sense together
Both clarify a concept and convey it.
My God, while well I know you think you're clever,

Your rampant ego now is overweening!
Thoughts so profound, their only shape is formless?
In form lies rhythm, metre, time and meaning.
When you reject the lot, what's left is gormless –

Like *pasta al tartufo* minus truffles
Or vintage Piper-Heidsieck less the bubbles.
The haunted sound of silent absence muffles
The whitespace racket from your tin-eared troubles.

While as for politics and your elision
Of wine with wonky dreams: was there a genie
In that decanter? Come – do you envision
Your life as London's proto-Mussolini?'

ORIGINS

With this exchange, the gods of mischief started
A viral avalanche from almost nothing.
Hell hath no rancour like the man outsmarted,
Trapped by the she-bear in his act of bluffing.

He should have bit his tongue and counted twenty,
Or swallowed down his pride and breathed out slower.
Instead, from his hubristic horn of plenty
Erupted fumes of human Krakatoa:

'You contradict me! What goddamned presumption
To interrupt with views so undiscerning.
Who made *you* the expert? Sheer brazen gumption
To foist on us your little twee-twee learning!

Just as the charms of youth have fled your body,
I fear your pretty mind is growing frumpy.'
'Robert-the-paunch,' she said, 'that shot was shoddy.
You feel a failure and it makes you grumpy.

Wealth, you once claimed, sets free the mind for culture.
Ten years a partner — how that dream has ended!
Your nightingale egg hatched a corporate vulture
Who feasts on bankrupts and whose sense descended

To fawn on have-yacht hedgies guzzling Margaux
As "gentlemen of taste", not moneyed winos.'
'You…cow! Your "style" resembles Mar-a-Lago,
Your ladies' luncheons – herds of pregnant rhinos!

Emotional mush and mwaa-mwaa on babies,
Pianos, nurseries and private tutors.
Eton or Radley? All this talk of maybes,
But none could pass the cut to work at Hooters.'

Spoke he. All froze. In shock, in total silence
She quit this room resounding with fiasco.
Returning tranquil, she with sudden violence
Flung in his face a pan of hot Tabasco.

'Enjoy your just dessert – of scorpion pepper,
And picked to match your taste: grotesque but famous!
You scumbag charlatan, you moral leper,
You asinine large-bottomed ignoramus,

Style is the feather guiding home the arrow,
Not peacock plumery in your fedora.
Silk-like and fragile, delicate yet narrow,
It sticks a sharpened point like La Camorra.'

ORIGINS

En garde! Touché! Slashing out, then recoiling,
The combatants unleash their tongues. Attack from
Him, now repulsed by her. Emotions boiling,
Considered thought has yet to haul them back from

Their fast crescendo of allegro madness.
In place of peace and harmony, their swarm of
Insults — dispatched in wrath, recalled in sadness —
Electrify this vicious thunderstorm of

Hatred. But spouses fight then make up — plucking
Out barbs with honeyed words; taking back; saying
Sorry, then kissing, stripping and wildly drinking mugs of hot cocoa.
They wash away woe with the wordless slaying

Of ugly argument and disputation.
Forgiveness will forget and not disparage
The contrite heart or seek its subjugation.
But, sad for them, still sadder for their marriage,

Sasha was listening: a social climber
With dreams of money, fame and recognition.
Alert to fortune, this inventive rhymer
Now saw and seized his moment. His ambition

GALAXIES

Was their demise: too soon, his thumbs had tweeted
The world with everything. As passions harden,
Their cruel jibes, retweeted and repeated,
Are sped from Ballarat to Baden-Baden.

They roar in Roma, laugh in Barcelona;
Nepali snows echo to giggling Gurkhas.
One smile infects one cop in Arizona
And Kandahari women smirk in burkas;

While those whom unemployment and recession
Has seized then thrown behind, betrayed and moody,
Find schadenfreude tempering depression
Through this aristo-farce of Punch and Judy.

Their teacup tempest, speedily arisen,
Now gusts throughout the nations: pointed, candid,
Caught in the interweb's electric prison,
Each slur preserved as fresh as first it landed.

This ding-dong spousal spat of moral failing –
Her brief if sharp outbursts of green-eyed fury,
His pompous chauvinistic me-me wailing –
Are all exposed naked before the jury

ORIGINS

Of every land and place and time and people.
While still in anger-land, their homely drama,
Whipped on by merciless fast-fingered tweeple,
Is magnified into a panorama

Of discord, bickering and malefaction.
It reaches everywhere; its stench is rotten;
And neither humble prayer nor libel action
Can cause it to deflate and lie forgotten.

When local hullabubs, while yet in fashion,
Flare up with viral growth beyond recalling,
They spread and replicate themselves; the flash in
A pan becomes a forest blaze, installing

Descendant copies of the kindling brouhaha.
These take its imprint, then diverge in pattern:
As phrases like *Avoir la pêche* or *Ouh la la!*
In time evolved their forms from Caesar's Latin.

So if you thought eternal noble glory
Shines out of galaxies, by now you know that
A humbler common birth commenced their story.
The laws of science and social networks show that

GALAXIES

Their quantum origins resemble Twitter:
Rough seas of storm-tossed noise, from which was chosen
One frothy splash amidst the foam. This flitter,
First amplified, inflated large then frozen,

Grew to a dreamy cosmic elven tower.
Like willows weeping diamond snow, its spirals
Bewilder lovesick eyes, infused with power
To sweep away our hearts within their chiral

Fantasy. Old or young, at home or college,
The galaxies call. Dumb in awe, eyes numb from
Their ghostly majesties, we have the knowledge
To answer all who ask, 'Where *did* they come from?'

Notes on *Elements*

p. 11 Bronze is not itself an element, but rather an alloy of the elements Copper and Tin. As technology and weaponry evolved from softer metals such as Copper through to Iron, so also the formation of elements in stars starts with lighter elements before culminating in Iron.

p. 11 Both stubby pencil leads and diamonds are made of Carbon atoms. However, the crystalline arrangement which characterises diamond can only form under conditions of extreme pressure and heat.

p. 11 The Standard Model of particle physics contains three forces (strong nuclear force, weak nuclear force and electromagnetism). The particles responsible for carrying these forces are the *gluon* for the strong force, the *W* and *Z* for the weak force and the *photon* for electromagnetism.

The W (strictly the W^+ and W^-) and Z particles are massive (each, individually, is heavier than an atom of iron). They are unstable and grossly obese cousins of the photon, the electromagnetic force carrier. Their mass arises through interactions with the *Higgs field* (whose excitations are the *Higgs boson*). The W and Z were discovered at CERN in 1983 by collaborations under the mercurial and tempestuous leadership of Carlo Rubbia, for whom the term alpha male suggests the Greek alphabet is missing a zeroth letter.

Given the domineering strength of the strong force in all subnuclear processes, the rest of the Standard Model particles are usefully grouped into those that do not feel the strong force (known as *leptons*) and those that do. The latter (which may in addition feel the weak and electromagnetic forces) are the *quarks*. Their different types – often called flavours – are the *up, down, strange, charm, bottom* (occasionally called *beauty*) and *top* (occasionally called *truth*).

The leptons feel at least one of the electromagnetic forces and the weak nuclear force. They are the *electron, muon* and *tau*, together with the corresponding *electron-neutrino, muon-neutrino* and *tau-neutrino*. These also have anti-particles (such as the *anti-muon* and *anti-tau*, although the *anti-electron* is known as the *positron*).

Most of the particles of the Standard Model are unstable and decay to other particles with lifetimes much less than a second. The observation and detection of particles occurs via the interaction of either particles or their decay products with normal matter. The *neutrini* are the hardest to detect; a sheet of lead as thick as the Earth–Moon distance would stop less than one neutrino in a million.

The golden age of the Standard Model lasted from 1967 to 1983. In this period, experimental physicists discovered the majority of the particles of the Standard Model while their theoretical colleagues wrote down equations for it that have survived two generations and fifty years' worth of attempts to prove them wrong. Bastards!

p. 12 Arsenic, Mercury, Polonium, Uranium – in some ways, the Periodic Table reads like an extract from a crime novel.

p. 12 Ancient alchemy and astrology are arguably the two most unfairly mocked endeavours in the history of science; the contemporary contempt for them is one of the many gifts of lazy prejudice that the worst minds of the present bestow on the best minds of the past.

There are many who deride as obvious nonsense the astrological idea that planetary arrangements in the sky affect human fortunes and emotions – yet think it boringly true that an invisible gravitational force emanates from the Moon two hundred and forty thousand miles away to cause the tides. Astrology was and is wrong; but the purgatory that is GCSE physics has dulled the bizarre and shocking nature of Newton's idea of universal, instantaneous, at-a-distance gravitation written in the language of mathematics. Likewise, the act of boiling an egg illustrates the remarkable textural changes that simple heat can deliver. Given this, why not think that some other combination of pressure, temperature and chemicals could turn one of Earth's many minerals into Gold?

p. 13 The atomic nucleus is, even by the standards of the atom, a tiny and concentrated ball of charge. While the typical size of an atom is one ten-billionth of a metre, corresponding to the orbital radii of the electrons, the size of the central nucleus is one hundred thousand times smaller, at one million-billionth of a metre (or one *femtometre*, written as 10^{-15} m in scientific notation).

p. 13 The mathematical forms for the strength of the gravitational force and electrostatic force are identical (both drop off *exactly* with the square of the distance). Why not one power of distance, or three, or one and a half? It is not a coincidence. For reasons that go beyond these notes, this equivalence is required by mathematical consistency.

p. 14 As well as his scientific achievements, Isaac Newton (1642–1727) was also Warden (1696) and then Master (1699) of the

NOTES ON *ELEMENTS*

Royal Mint. The care he took over practical measurements to ensure the integrity of the King's currency was matched by that lavished on the souls of those who would counterfeit it, by ensuring a rapid reunion with their Creator via Tyburn.

pp. 14–15 Most isotopes are unstable; their ratio of *neutrons* to *protons* is either too large or too small. Radioactive decay processes (either alpha- or -beta decays) change these ratios and pull the isotopes towards a small narrow belt which charts the stable isotopes on the graph of neutron number versus proton number: a belt known as the *valley of stability*.

pp. 15–16 The proton (electric charge +1) is a composite particle made from two *up* quarks (each charge $+2/3$) and one *down* quark (charge $-1/3$). However, the inner life of the proton is set not just by these quarks alone but also by the internal mutual dynamics between them.

It is one of the trickiest problems in the quantum theory of the strong nuclear force to obtain a precise description of this inner life. The internal 'stuff' of the proton is divided into *valence quarks* (the two ups and one down mentioned above) and *sea quarks*.

The *sea quarks* have no permanent existence and are also called *virtual* quarks. Using quantum mechanics, they briefly 'borrow' energy from the vacuum and can only exist via this borrowed energy. A consequence of this is that less massive quarks (which, via $E=mc^2$, need less energy in order to exist) can borrow energy more easily. Therefore, the most common *sea quarks* are the lightest *up* and *down* quarks, followed by the *strange* quark, and then the *charm* and *bottom* quarks. The ultra-massive *top* quark is, effectively, never found as a *sea quark*.

Also found in this sea is the *gluon*, the force carrier of the strong interaction, either individually or as a collective of many *gluons* (called *glue*). For deeply technical reasons, the more energetic (or *harder*) a proton's interaction is, the more *gluons* are found in the sea.

The above sounds esoteric but is well supported by experiments using the time-honoured techniques of toddlers and collider physicists: smash stuff together and gaze with joy and amazement at the debris.

p. 16 *Legs* and *loops* are two of the basic ingredients of Feynman diagrams – the cartoon-like squiggles invented by the American Nobel Laureate Richard Feynman (1918–1988) to simplify particle physics calculations, 'bringing computation to the masses' as his friend and rival Julian Schwinger said.

p. 17 The energy required to dissociate the subnuclear bonds of the proton is over a million times greater than that required to dissociate the inter-atomic bonds that hold molecules together.

p. 18 The *electrons*, *positrons* and *muons* present in the early universe are all *leptons* and do not feel the strong force; in contrast to the *neutrons*, *pions* (pairs of up or down quarks and antiquarks, with all combinations possible) and *kaons* (either a *strange* quark paired with an *up* or *down* antiquark, or a *strange* antiquark paired with an *up* or *down* quark).

p. 18 Hydrogen, simplest of all elements, has a nucleus consisting of a single proton, in turn orbited by a single electron (every other element also has at least one neutron in its nucleus). *Ionised* elements have some or all of their electrons removed; for Hydrogen, ionised Hydrogen is therefore equivalent to the residual single proton of its nucleus.

p. 18 This identification of Venus belongs to prehistory and was well known in the ancient world: 'stella Veneris, quae *Φωσφόρος* Graece Lucifer Latine dicitur cum antegreditur solem, cum subsequitur autem *"Εσπερος*' ('the star of Venus, which is called Phosphorus in Greek and Lucifer in Latin when it precedes the sun, but Hesperus when it comes after the sun'), Cicero, *de Natura Deorum* 2.53.

pp. 18–19 No free quarks exist now as independent particles; today, they are only found inside the proton (or similar particles such as the neutron) and cannot be liberated from it. These quarks exist within bound states; either as a throuple with two other quarks (called a *baryon*) or as a quark-antiquark pair (called a *meson*). This absence of free quarks is one of the most striking differences between the strong force (which all quarks feel) and electromagnetism. For example, free *electrons* are commonplace, which enabled the electron's discovery first as a particle by J. J. Thomson in 1897 and subsequently as a wave by his son George Thomson.

p. 19 The note to the last note is that, in the extremely early universe, free quarks (and free gluons) *did* exist. The high temperatures (over a trillion degrees) present made the universe then qualitatively different to the universe now. This early phase lasted ten millionths of a second, until – like water cooling and freezing to ice – quarks joined into composite objects (such as the proton) and the story of the Elements truly began.

p. 19 *Valency* refers to an atom's capacity to form chemical bonds to other atoms.

p. 19 Hydrogen, the workhorse of chemistry, will bind to almost any other element in the formation of chemical molecules. The series of *ethane* molecules, which starts with Methane and

Ethane, is described by chains of Carbon atoms with attached Hydrogen filling in all the other bonds.

p. 20 Although the neutron (one *up* and two *down* quarks) has no electric charge, its mass only differs from the proton (two *up* and one *down* quarks) by 0.15 per cent. This great commonality, also seen in the way the proton and neutron interact with other particles, arises as the pair are identical with respect to the strong force.

pp. 20–21 Water becomes heavy water when its Hydrogen atoms are replaced by Deuterium. The resulting change in mass affects the bonding properties of water molecules and subtly disrupts the biochemical equilibrium of the body, producing symptoms analogous to radiation poisoning. Heavy water becomes fatally toxic to mammals when its concentration reaches around twenty per cent of the total water content in the body.

pp. 21–22 Although the *deuteron*, the combination of a proton and neutron, is stable – an isolated deuteron will never decay – it only needs a small amount of energy (where 'small' is in comparison to typical nuclear energies) to break it up, like a house of cards that can be easily knocked apart. In contrast, the *Helium* nucleus (two protons and two neutrons) is a robust and stable configuration. Mathematically, the reason for the stability of the Helium nucleus is very similar to the reason for the stability of the Helium atom, the inert element arising from the Helium nucleus bound to two electrons.

pp. 22–23 To appreciate the jaw-dropping differences in the relative strength of the fundamental forces, realise that your little finger, using electromagnetic signals that flash through nerve impulses to drive muscle contraction, can lift an object against the downward gravitational pull of all thousand billion billion tonnes of the Earth. Inside the nucleus, however, this same electromagnetic force is easily overcome by the strong force.

As all nuclei above Hydrogen contain multiple positive charges (their protons), the difficulty in forming such nuclei lies in bringing these positive charges, all of whom repel each other, close enough that the attractive effects of the strong force can take over.

p. 23 The empirical behaviour of the strong force is subtle and complex. It is absolutely dominant at nuclear and subnuclear distances but falls rapidly to irrelevance outside the nucleus. For decades, this odd behaviour appeared mathematically inexplicable and belonging to the *hic sunt dracones* part of the map of physics. In 1973, the inexplicable became explicable and, three decades later, the 2004 Physics Nobel Prize would be awarded to David Gross, David Politzer and Franck Wilczek for their work on understanding the strong nuclear force.

pp. 24–25 Charles-Augustin de Coulomb (1736–1806) was an engineering officer in the French army. Governed by the needs of his employer, his more practical work focused on soil mechanics and the retaining walls that are a necessary part of the construction of military forts. On a more fundamental level, his eponymous law of electrostatics describes how the repulsive (attractive) force between two same (opposite) charges falls off with distance: that is, with the same inverse square law as for gravity.

pp. 27–30 The formation of the lightest elements in the early universe is called *Big Bang Nucleosynthesis*. This epoch starts when the universe is approximately one second old and terminates approximately ten minutes later. During this epoch, the universe is expanding and cooling: as it cools, the universe is no longer able to sustain the necessary levels of heat required to drive the nuclear reactions that would form more elements.

Considered as a hypothesis, Big Bang Nucleosynthesis makes distinctive predictions for the primordial ratios of the three lightest elements, Hydrogen, Helium and Lithium, and their isotopes. We say with confidence that Big Bang Nucleosynthesis occurred because these predicted ratios are a very good fit to observations, made through either the spectra of very old stars and nearby primordial gas, or from the properties of relic light from the early universe (the Cosmic Microwave Background).

p. 31 The precise time of formation of the first stars – the element factories of our universe – is not yet known precisely, but it is unlikely to be much earlier than one hundred million years. One of the principal goals of NASA's recently launched James Webb Space Telescope is to pin down this time and understand this epoch precisely.

Stars exist in equilibrium between an outward pressure originating in nuclear fusion and an inward pull that tends to gravitational collapse. Throughout their lifetime, their size and temperature evolve in the ways necessary to maintain this equilibrium. For more massive stars, the gravitational pull is always more intense and so they need to burn fiercer and brighter to balance it, resulting in a much shorter overall lifetime.

p. 32 The Matthew principle can be found in Matthew 25:29.

pp. 35–43 Cecilia Payne was born in Victorian London in 1900 and died in decidedly non-Victorian Massachusetts in 1979. Her PhD thesis, demonstrating that the Sun's chemical makeup is overwhelmingly that of Hydrogen and Helium, has been called 'the most brilliant thesis in the history of astronomy'. Payne was the name under which her biggest contributions were made and so I use it here (her surname became double-barrelled when she married, in 1934, the Russian astronomer Sergei Gaposchkin). The account here draws closely on her autobiography *The Dyer's Hand*.

p. 36 While at St Paul's Girls, the young Cecilia was privileged to hear one of the first ever performances of Holst's *Planets* in the school hall.

Obtaining a scholarship to Newnham College allowed her to attend Cambridge; the family were not poor, but neither were they wealthy enough that she could have attended Cambridge without some form of scholarship. Newnham and Girton were the two women's colleges, founded in 1871 and 1869 respectively, although it would not be until 1948 that Cambridge allowed women to receive actual degrees.

p. 37 Albert Charles Seward (1863–1941) was the Professor of Botany from 1906 to 1936. The expression 'Neither encouragement nor inspiration' is taken directly from Payne's own account of this meeting in her autobiography.

p. 37–38 Trinity College, Cambridge, was founded by Henry VIII and endowed with Henry's ill-gotten riches from the dissolution of the monasteries.

p. 38 Arthur Eddington (1882–1944) was the Plumian Professor of Astronomy at Cambridge and had obtained international fame from his leadership of the 1919 eclipse expedition which provided the first observational evidence for Einstein's theory of general relativity. He was a populariser of science before such activities became popular and gave many talks expounding Einstein's new theory of relativity to a wider audience.

For many years the most influential astronomer in the world, he appears here as part of the stories of both Cecilia Payne and Subrahmanyan Chandrasekhar.

p. 38 Ernest Rutherford (1871–1937, 1908 Nobel Prize) was one of the great experimental physicists of the twentieth century, responsible for discovering the presence of the nucleus at the centre of atoms.

Originally a farm boy from New Zealand, his research career had moved between Cambridge, McGill (Canada) and Manchester before he returned again to Cambridge in 1919 as a senior professor. Strong of build and strong of personality, he possessed a voice like a battleship's foghorn and epitomised the 'string-and-sealing-wax' era of nuclear physics.

p. 38 'Bosom' and its dual meanings: during a passage in her autobiography discussing love, Cecilia Payne recounts that 'it was 15 years before I outgrew my childish dream of playing the Beggar Maid to Eddington's King Cophetua'. This refers to a tale in which a powerful king with no previous interest in women sees a beggar girl and then immediately falls in love with her and proposes marriage.

As far as is known, throughout his life Eddington never showed any sexual interest in women (or men, for that matter).

pp. 38–39 The expression 'no insuperable objection' is also a verbatim quote reported in Cecilia Payne's autobiography.

William Smart was the technical assistant at the Cambridge observatory and the 'measuring machine' was a piece of astronomical instrumentation. Although there were better and simpler ways of doing the required calculation, the young Payne used the old original five-volume German text of the mathematician Carl Friedrich Gauss (1777–1855), which she obtained from the British library, to learn the method of least squares.

p. 39 The *Monthly Notices of the Royal Astronomical Society* is one of the oldest astronomical journals in the world. This paper can be found in Volume 83, Issue 5, March 1923, pages 334–338, authored by Miss Cecilia H. Payne and communicated by Professor A. S. Eddington FRS.

Such language is archaic: under modern publishing conventions, such a paper would most likely appear with Payne as first (or primary) author and Eddington as last (or senior) author.

p. 40 A set of particular historical circumstances resulted in many of the early important female astronomers coming from or through the Harvard observatory (among them Williamina Fleming, Henrietta Leavitt and Annie Jump Cannon). The senior astronomers at the observatory employed these women directly, on relatively low pay, in what they viewed as a menial task of classification. The lack of other employment opportunities meant that, despite the low pay, bright and intellectually curious women still took these jobs. And, yes, the job was classification: but what a dataset! They spent their hours with the largest and best dataset of stellar spectra in the world at the time, produced by one of the most advanced telescopes in the world; these women were immersed in all practical aspects of these spectra and made the discoveries to match.

p. 41 Meghnad Saha (1893–1956) was an Indian astrophysicist whose *Saha equation* relates the degree of ionisation, and thus the spectral properties, of gases to their temperature. Payne's great achievement came from using the vast Harvard dataset to test this equation and use the results to determine the temperature and composition of stars.

p. 41 A later student, Owen Gingerich, recalled that, 'A pack of cigarettes and a single match could get her through the entire [lecture] period.'

p. 42 Henry Norris Russell (1877–1957) was the director of the Princeton University Observatory and also the Russell behind the Hertzsprung-Russell diagram of stellar history and evolution.

p. 42 'Eppur si muove' ('But, yet, it moves') were Galileo's alleged words after being compelled by the Inquisition to recant the doctrine that the Earth moved round the Sun.

pp. 45–47 Fred Hoyle (1915–2001) was an archetype of that highly endangered species: the very British boffin. His background was typical of many of his generation: a family of modest means who valued education; the only one of his class to pass the 11+; attendance at a provincial grammar school; a scholarship to Cambridge and a subsequent brilliant career.

Hoyle was pugnacious and brilliant but never won the Nobel Prize. His collaborator William ('Willy') Fowler (1911–1995) was a less pugnacious, less brilliant scientist who shared the 1983 Nobel Prize with Chandrasekhar. Although the exclusion of Hoyle was surprising to many, the Nobel archives for that year are still sealed and no definitive explanation can yet be given. However, it surely would not have helped Hoyle that, among other scientific quarrels, he had previously impugned the integrity first of fellow British astronomer Antony Hewish (1974 Nobel Prize) and then of the Nobel Prize committee who had awarded Hewish the prize.

p. 46 The original collaboration between Hoyle and Fowler extended to include Geoffrey and Margaret Burbidge; between the four of them, they produced a definitive review (Burbidge,

Burbidge, Fowler and Hoyle) that became known to astronomers simply as B^2FH.

p. 47 One of the aphorisms attributed to Hoyle is that 'It is better to be interesting but wrong, rather than boring but right.'

p. 48 The Hertzsprung-Russell plot is the standard plot of stellar evolution used to illustrate the life cycle of stars, named after Henry Norris Russell and Ejnar Hertzsprung (1873–1967).

p. 48 In terms of stars, a *white dwarf* refers to a lighter star (with mass up to 1.4 solar masses, the Chandrasekhar limit) that has come to the end of its life. Such a star has ceased to generate heat via nuclear fusion and is now gradually cooling down; over a period of billions of years it will gradually turn into first a *red dwarf* and then ultimately a *black dwarf* as it cools and its colour changes. Such white dwarf stars mostly consist of Carbon as they were never hot enough to fuse any of the elements beyond Carbon.

pp. 49–50 A *Type IA Supernova* occurs when a light star below the Chandrasekhar mass limit accretes mass from some other source, thereby increasing its own mass beyond this limit. This immediately renders the star unstable and triggers gravitational collapse: the star falls in on itself and then explodes outwards as a supernova.

pp. 50–54 Subrahmanyan Chandrasekhar (1910–1995, 1983 Nobel Prize), widely known as Chandra, was one of the great astrophysicists of the twentieth century. Born into an atmosphere of learning (his uncle Chandrasekhara Raman won the 1930 Physics Nobel Prize), he first studied at Presidency College, Madras, before coming to England, where he became a Fellow of Trinity College, Cambridge. After 'events' (see below), he moved to Chicago and stayed there for the rest of his career.

His scientific style was famous and unique. Often the first into the department each morning, every seven or eight years he would pick a new topic. Through deep reading and calculation, he would master this topic completely and write a textbook on it. Having done this, he chose a different area and the cycle started again. The poem dramatises the events of a famous January 1935 meeting of the Royal Astronomical Society (RAS). Chandra had been working for some time on his (correct) account of how massive stars cannot be supported by quantum (Fermi) pressure and he was due to present this work at the meeting. To Chandra's surprise, Eddington inserted himself onto the schedule immediately after Chandra (Chandra had no idea that Eddington was working on any similar topic). However, when Eddington came to talk, he simply dismissed Chandra's ideas and stated, to laughter from the room, that they were physically nonsensical. The quoted dialogue is invented; the impact is not.

p. 51 A *bra* and a *ket* are the names of the mathematical quantities that describe the state of a quantum-mechanical system; they were coined by the famously taciturn Paul Dirac.

There is a story about Dirac at the high table in his Cambridge college, St John's. Dirac was eating in his customary silence while his humanities colleague discoursed broadly about language, words and their etymologies, plus any neologisms they had introduced into the language. Dirac opened his mouth. 'I invented the bra,' he said – and returned to his own thoughts.

p. 51 'Professor Newton tries to collapse the star; Professor Fermi tries to hold it up; Professor Chandrasekhar is the adjudicator' – a concise summary I heard as an undergraduate but whose original author is unknown (to me).

NOTES ON *ELEMENTS*

p. 55 Although not visible to the naked eye, the bright surface of a star contains many small dark sunspots. In addition, stars are surrounded by a *corona* of magnetised gas, much hotter than the surface, which sits a little bit above the surface and shines brightly in X-rays. As the surface temperature of stars is not hot enough to emit X-rays, the X-ray view of a star is of a bright ring surrounding a dull disc.

p. 57 Hannes Alfvén (1908–1995) was a Swedish engineer, winner of the 1970 Nobel Prize, who pioneered the study of hot magnetised plasmas and the propagating waves that could exist within them.

p. 58 Chlorine was the principal component of the 'Gas! GAS! Quick, boys!' shells of World War I.

Pure metallic Potassium is extremely soft at room temperature (although it will rapidly react with oxygen if exposed to the atmosphere).

pp. 59–60 Iron is the most stable of all elements. For all elements lighter than Iron, it is energetically favourable for them to merge up the Periodic Table: but for those heavier than Iron, energetics prefers them to fission and move down the Periodic Table. The elements up to Iron are formed inside stars in the course of their lifetime; the elements beyond Iron are formed when a star explodes (or, later in the universe, when two stars collide).

pp. 61–63 We expect to observe supernovae (exploding stars) from inside our own galaxy approximately once per century (the next one is somewhat overdue and is the most eagerly awaited event in all astronomy). Supernova 1987A was the most recent 'close' supernova, occurring not in the Milky Way but in the Large Magellanic Cloud, a nearby satellite dwarf galaxy of the Milky Way around 150,000 light years away.

NOTES ON *ELEMENTS*

Type II supernovae arise from the collapse of massive stars at the end of their lifetime under their own gravity; one of their most distinctive signatures is the neutrino burst that occurs within the very last moments as the densities reach a point where the weak interaction (which is normally completely negligible compared to the strong and electromagnetic interactions) becomes important. In contrast to SN1987A, the next galactic supernova (when it occurs) is expected to result in tens of thousands of detected neutrini.

Viewed through telescopes, these supernovae leave an expanding ring on the sky. One of the most famous of these is the *Crab Nebula*, the remnant of a 1054 supernova recorded in Chinese (but not European) sources. The enormous energies and densities released in these supernovae force nuclei into highly radioactive agglomerations of neutrons, which subsequently turn into stable heavy nuclei via long chains of radioactive decays.

pp. 63–65 The poem glosses over some subtleties to do with the formation of the heaviest elements (i.e. those such as Uranium). True: the heaviest elements are formed by a dense hail of neutrons. As described in the poem, this process (whose technical name is the *r-process*) forms neutron-rich agglomerates which subsequently produce, via their radioactive decays, the heaviest stable elements.

What is glossed over is the fact that there are two possible ways the r-process can occur, both associated with exploding stars, and it is not yet fully known how each contributes to the formation of heavy elements in the present universe. One way is in supernovae from massive stars (as described in the poem); the other (not described in the poem) is in the later destruction of neutron stars through their collision with either black holes or other neutron stars. Both ways are likely to be operative in the present universe, although their relative importance is unclear.

However, neutron star collisions do require as a necessary pre-requisite the prior existence of neutron stars. Neutron stars are always formed as supernova remnants; depending on the precise properties of the supernova, its core ends up either as a black hole or as a neutron star. In the history of the universe, the supernovae come first; the neutron stars later. Today, both supernovae and neutron star collisions are observed as regular (on cosmological scales) events in the universe.

For the reason above, the poem focuses only on supernovae in its description of the original formation of the heaviest elements. The first generation of stars, called *Population III stars*, almost all exploded as supernovae early in the universe's history. Before they exploded, there were no neutron stars anywhere. As long as the r-process operated at any level in these first supernovae of Population III stars, it is here we find the first origins of the heaviest elements.

p. 71–72 Paracelsus (c.1493–1541) was one of the Renaissance figures at the intersection of medicine, astronomy and astrology, a combination which now seems incomprehensible. Ibn Rushd (1126–1198, also known in the West as Averroes) and al-Khwarizmi (c.780–c.850) were both Islamic polymaths for whom astronomy was just one of their many interests; the latter's studies of mathematics have given us the word 'algorithm' as a corruption of his name. Tycho Brahe (1546–1601) was a Danish astronomer and the last great observer prior to the invention of the telescope; he ended his life as the court astronomer to Rudolf II, the Holy Roman Emperor. Enrico

Fermi (1901–1954) was an Italian physicist often regarded as the last universal physicist, being equally competent in both experimental and theoretical physics. The quantum Fermi pressure that can support dead stars is named after him. His early death from stomach cancer was probably a consequence of his work on constructing early nuclear reactors as part of the Manhattan Project. Richard Feynman (1918–1988) and Steven Weinberg (1933–2021) were giants of twentieth century American physics, who both played key roles in the establishment of the Standard Model of particle physics.

The James Webb Space Telescope, named after a distinguished former head of NASA, is the latest and best version of that first generation of telescopes Galileo knew. We are familiar with physical courage; there is also a type of moral courage, where nations invest thirty years and ten billion dollars to mount delicate electronics and precision optics on a tube of rocket fuel in the hope that nothing will go wrong on ignition.

p. 72   The Barley Mow is a pub in Clifton Hampden, Oxfordshire.

Notes on *Galaxies*

p. 78   Thomas Edison (1847–1931) and Joseph Swan (1828–1914) were respectively the American and English inventors of the lightbulb; each held the patent in his own country. Their combined commercial force resulted in the Ediswan bulb, whose flickering filaments would enlighten Sherlock Holmes, Dr Watson and other inhabitants of the late Victorian world.

p. 79   Ptolemy's *Almagest*, together with its geocentric Aristotelian worldview, represented the basis of Western astronomy until the time of Copernicus.

So much history in two words! The originally Homeric name of this Alexandrian polymath (c.100–170) reflects Greek inculturation in Egypt; the name of his text (a corruption of *al-megistos*, 'the greatest') reflects the transmission route of classical culture from Greek to Arabic to mediaeval Latin.

p. 79   Those seeking a model of how not to win political support for an idea can take Galileo (1564–1642) as their example. To convince the world of the heliocentric universe, he wrote, in the Italian vernacular, his *Dialogue Concerning the Two Chief World Systems* (1632) and placed all of the Pope's favourite (and wrong) arguments for a geocentric universe in the voice of a character called Simplicio.

Of course, Galileo was right: but in just as much trouble as if he'd been wrong.

Galileo's earlier adoption of the telescope (building on work by Dutch glassworkers) led to his 1610 book *Siderius Nuncius*

('Starry Messenger'), notable as the first astronomy text based on observations with a telescope. In it, he describes his observations of the moons of Jupiter, which were unknown to classical astronomy. These mark the transition from the old astronomy to the new, with the realisation that there was, indeed, plenty new under the sun.

p. 80 Mount Palomar (in California), Mauna Kea (in Hawaii) and Siding Spring (in New South Wales, Australia) are three elevated mountain sites of both historic and continuing importance for astronomical observations that rely on clear skies and no light pollution.

pp. 80–81 *Quasars* (a contraction of quasi-stellar radio sources) were first discovered in the late 1950s, as compact, energetic sources of radio waves with (then) no corresponding optical counterpart.

A mystery for many years, they are now known to arise from supermassive black holes swallowing gas, at cosmological distances far, far beyond the Milky Way: an identification which showed that the universe was, once again, much larger and grander than previously imagined.

pp. 81–82 The large-scale isotropy and homogeneity of the universe, sometimes called the Copernican principle, is arguably the single most important and striking observational feature of the universe: on large scales, the universe looks the same in all directions (on the smallest scales, children learn rapidly that down is a preferred direction and that objects do not fall upwards).

p. 84 Q. E. D. – *quod erat demonstrandum* '(the result) which was to be shown' – was familiar to generations of schoolchildren at the end of every proof inside their battered copy of Euclid's *Elements*.

p. 84 In a universe where everything emerges from a Big Bang, it sounds really very surprising that there could be parts of the universe which have never had any causal contact with each other: how can this be possible if we start with a singularity from which everything emerges?

Although counter-intuitive, this result does follow from the mathematics, where the easiest way to see it comes from a particular choice of coordinates due to the British mathematician Roger Penrose (1931–, 2020 Nobel Prize).

This fact – that the 'ordinary' Big Bang cosmology provides no explanation for the homogeneity and isotropy of the universe – is a key reason why there must be something extra in the early universe.

p. 85 Viatcheslav Mukhanov (1956–), Paul Steinhardt (1952–), Andrei Linde (1948–), Alan Guth (1947–) and Alexei Starobinsky (1948–2023) were all pioneers of inflationary theory. Although the first stirrings of these ideas occurred in the late 1970s, the most clearly defined start to the subject is Alan Guth's paper on the subject published in a January 1981 edition of *Physical Review D*.

Guth (and to some extent Starobinsky) get the credit for the original proposal of an inflationary epoch in the early universe and the realisation that a period of cosmic inflation would make the universe homogeneous and isotropic on very large scales.

The contributions of Linde, Mukhanov and Steinhardt were to realise that inflation did not only explain why the universe was almost perfectly homogeneous on large scales: it also provided a mechanism, through quantum mechanics, to generate small inhomogeneities that could grow under gravity into stars and galaxies. From a modern perspective, it is this latter aspect, the ability to explain the origin of structure in the universe, that is regarded as the biggest success of inflation.

Linde, Mukhanov and Starobinsky were all products of the Soviet school of theoretical physics: a school with a distinctive style and culture but now receding into history. The pupils went west; the teachers stayed and grew old, and none came to replace them.

pp. 86–90 The discovery of general relativity, the culmination of a decade of work starting in 1905, was the greatest of many great achievements by Albert Einstein (1879–1955). Einstein represents, for good or ill, the popular icon of what a theoretical physicist is.

p. 88 The *equivalence principle* is one of the core concepts of general relativity. It represents the idea that no purely local measurement can ever distinguish between a constant gravitational pull and a uniform background acceleration.

p. 88 Bernhard Riemann (1826–1866) was one of the leading mathematicians of the nineteenth century, famous for his work on understanding and quantifying curved geometries. His eponymous *Riemann tensor* is an integral part of all mathematical formulations of Einstein's theory.

A *manifold* is the mathematical term for a geometric space (modulo some distinctions that do not belong in these notes).

p. 89 What the mathematician Hermann Minkowski said about Einstein's earlier theory of special relativity is even more true of his general theory: 'Henceforth, space of itself and time of itself will disappear into mere shadows, and only a kind of union of the two will survive.'

pp. 89–90 The Einstein equations are written as $G_{\mu\nu} = 8\pi G\, T_{\mu\nu}$, where G denotes Newton's gravitational constant; the notation encompasses ten individual equations. To many physicists, these are not only the most beautiful equations in the entire subject but also one of the most beautiful objects in all human history. On the left-hand side is the *Einstein tensor*, an object constructed from the Riemann tensor and encapsulating the spacetime geometry. On the right-hand side is the *stress-energy tensor*, which tells us about all the conventional matter, particles and energy present (for vacuum solutions, this right-hand side is absent).

Another feature of the elegant simplicity of Einstein's theory is the way objects move through spacetime on *geodesics* of the geometry. Analogous to the great circles of Earth, these paths are locally straight and minimise the distance between two points.

Michelangelo (1475–1564) may have been a genius, but his work is stamped with his particular High Renaissance era and culture. In contrast, any civilisation, whether human or alien, that thinks seriously on gravity and space and time must come upon Einstein's equations. In alternative histories of the Earth, the Sistine Chapel never existed: but Einstein's equations do, and these equations were discovered in Mumbai or Mombasa with exactly the same form as found by Einstein.

p. 91 Sagittarius A* was originally the name of an unexplained radio source in the constellation Sagittarius. It now refers to the explanation of the source: the presence of a supermassive black hole at that point which defines the centre of the Milky Way. The stars of the Milky Way orbit around this black hole, whose immediate vicinity is filled with a very high density of stars, gas and dust whose combination gives rise to the observed radio emission.

p. 92 The complexity of the vacuum is one of the recurring themes of both particle physics and general relativity across the last hundred years. It is hard to overstate how subtle, both physically and mathematically, the concept of 'nothing' is.

The idea that there could be an energy associated to this 'nothing', an energy associated to the vacuum of empty space, violated Einstein's sense of aesthetics and he originally forbade it on grounds that were little more than arbitrary. Homer nods and Einstein dozes: however, Einstein subsequently recognised that he was just imposing his prejudices on something far more subtle, calling this his biggest mistake.

pp. 92–94 The inflationary expansion of the universe is not that of your grandparents' Big Bang, however staggering even that may appear. Although the inflationary epoch is brief when measured in seconds, during it the growth of the universe is overwhelmingly more rapid than in any other epoch. The quantitative way of putting this is that the expansion is exponential in time rather than power-law.

Those who like powers of ten may note that, during inflation, the universe may have increased in size by a factor of over 10^{24} (one million billion billion, approximately eighty doubling factors) in a timescale as short as 10^{-36} seconds (a billionth of a billionth of a billionth of a billionth of a second).

p. 94 Lyonesse is a legendary land between Cornwall and the Isles of Scilly that was swallowed up by the ocean in a single night: Atlantis with red buses and cups of milky tea.

p. 95 Threadneedle Street is in London and is the site of the Bank of England; any relevance to Matthew 19:24 is left to the reader.

pp. 96–98 Of all the appealing aspects of physics, one of the most appealing is the way it links the very large to the very small. This is not just true of cosmological inflation and the large-scale structure of the universe; it also applies to our knowledge of the components of distant galaxies, derived from quantum-mechanical atomic emission lines from the chemical elements present in them. To understand the big, look to the small.

p. 100–101 It is only when working on the scale of atoms, at around a tenth of a nanometre, that the quantum nature of the world becomes unavoidable: it is here the classical Newtonian picture breaks down irrevocably. We are creatures whose evolutionary instincts come from the larger scales associated to eating and being eaten; our minds struggle to interpret quantum mechanics as we ponder the question of what quantum reality ultimately represents.

For young physicists, this period of reflection can be regarded as a spell in a juvenile detention centre: a soul-shaking period of six to twelve months redeemed through a useful later life of rectitude in prediction and calculation. Philosophers of science, however, may sometimes find themselves sentenced to life imprisonment.

p. 102 The wavefunction is the central object in quantum mechanics; if we know the wavefunction, we know the configuration of a quantum system. By convention, the wavefunction is most often denoted by the Greek letters *phi* or *psi*.

pp. 102–107 Werner Heisenberg (1901–1976, 1932 Nobel Prize) was one of the great physicists of the twentieth century, right at the heart of the Knabenphysik of the 1920s that toppled several centuries of wisdom. His life encapsulates the central enigma of the twentieth century: how did a society that, in many ways, led the world in culture and science end up focusing much of its efforts on the systematic massacre and elimination of the Jewish people? On an individual level, Heisenberg looked down on the Nazis – but, nonetheless, he spent the Second World War leading their atomic bomb project.

p. 103 LMU, Ludwig-Maximilians-Universität, is located in Munich, and Heisenberg's father, August Heisenberg, was from 1910 the Professor of Byzantine Studies there. It is the premier university in Bavaria and one of Europe's leading research universities. At one point as a student during the German civil war immediately after the First World War, the young Heisenberg would sneak through the front lines between the Whites and Reds to fetch food for his family from a farm just outside Munich. Later in life, as a distinguished professor and Nobel Laureate, he constructed his Max-Planck-Institut on the same location, next to two gardens: Englischer- and Bier-.

p. 103 Arnold Sommerfeld (1868–1951) was the elder statesman of the German quantum physicists, supervising seven future Nobel Laureates; Niels Bohr (1885–1962, 1922 Nobel Prize) played the same role for European physicists as a whole.

p. 104 Helgoland (also known as Heligoland) is a small island in the North Sea free from pollen. It was here, late one night, that the structure of quantum mechanics became clear to Heisenberg; the account given in the poem, including the dawn walk, is based on Heisenberg's own account.

We can all take either inspiration or fright from Heisenberg's words a few months earlier: 'I have not thought at all about Physics in the last month, and I don't know if I still understand anything of it.'

p. 104 Wolfgang Pauli (1900–1958, 1945 Nobel Prize), Erwin Schrödinger (1887–1961, 1933 Nobel Prize), Paul Dirac (1902–1984, 1933 Nobel Prize) and Max Born (1882–1970, 1954 Nobel Prize) were all pioneers of quantum mechanics. With the exception of Dirac, the discovery of quantum mechanics emerged almost entirely from Germanic (broadly understood) university culture.

p. 106 *Die Fahne hoch* ('Raise the flag') are the opening words of the Horst-Wessel-Lied, the anthem of the Nazi party, which was sung at many official events (it has been made illegal in modern Germany). German academics are civil servants; from 1934 a personal loyalty oath to Adolf Hitler as Führer became required of all civil servants, replacing a previous oath to the German constitution.

The Deutsche Physik movement had as its goal the creation of a movement of national physics free from Jewish influence; it became (in)famous for its attempt to reject (the Jewish) Einstein's theories of relativity. The movement was behind the 1931 publication *One Hundred Authors against Einstein* (to which Einstein responded, 'Why a hundred? If I were wrong, one would have been enough.').

p. 107 The precise statement of Heisenberg's uncertainty principle is that the product of the uncertainty in our knowledge of a particle's position with the uncertainty in our knowledge of its momentum must always exceed a fixed constant (Planck's constant). In simple cases, the momentum of a particle equals the product of its mass and its velocity. In mathematical language:

$$\Delta x \; \Delta p \geq \hbar / 2$$

This uncertainty principle is, via some mathematical trickery, what makes the vacuum so complex to understand, as it implies the impossibility of ever resting in a fixed position with zero motion.

p. 109 Time plays a slightly more fundamental role than space in the laws of physics. We know of toy models where our notion of space arises as a quantitative approximation to something else that is more fundamental; but not, as yet, for time.

p. 110 The gas which lies between the different galaxies inside a galaxy cluster is at a density of less than one particle per cubic centimetre. Nonetheless, the typical temperature of this gas is hundreds of millions of degrees (this fact is slightly shocking when one first learns it; I know, by experience, it shocks bright Oxford physics undergraduates well into their course). As with any fire (or oven), this hot gas emits light: at such a temperature, the light is in the form of X-rays. Through these X-rays, we understand the form and structure of galaxy clusters.

pp. 110–111 The probabilistic nature of quantum mechanics is one of its more counter-intuitive features; we cannot say what *will* happen, only what *probably will* happen. The mathematics of the subject delineates the precise ways these probabilities appear and are calculated.

p. 113 The *metric* is the mathematical object that describes the geometry of spacetime; Python is a commonly used programming language.

pp. 113–115 The theory of cosmic inflation tells a remarkable story of microscopic quantum fluctuations which are stretched across the universe and then grow into galaxies. The most remarkable feature of all is that this story appears highly likely to be true: the detailed predictions of cosmic inflation are all borne out in observations of the Cosmic Microwave Background, the relic light which fills the universe and originated a few hundred thousand years after the Big Bang.

Although it remains technically possible that some other theory might be able to replicate the achievements of cosmic inflation, its combination of simplicity, success and explanatory power makes the inflationary theory firmly established as the standard account of the very early universe.

Acknowledgements

From the first squawl, my parents – by example they taught me that enjoying the two cultures is as natural as breathing with both lungs. Their blend of encouragement and open bribery led the child me to learn poems by heart and treasure their soundscape.

From school, Boniface Moran OSB, who first showed me the grandeur of epic poetry and what it could do.

To all my colleagues in physics, past, present and future: you have rewritten the world. Your work inspires my attempts here to describe these best ideas with the best words in the best order. Where I have failed, let others go three steps better.

In Oxford – one of the special geniuses of this university lies in the colleagues it provides – I thank Craig Raine, Will Poole and Hannah Sullivan for reading drafts of this book at various stages and giving me confidence in my own ability to write it. Another is the quality of the students one gets to teach. Among them, I thank Grace Haworth for reading a draft version and providing feedback.

At Wylie, my agent, Luke Ingram, for explaining the publishing industry to me, challenging me to refine and improve the text, and packaging the work with a large red ribbon to send to publishers.

Everyone at Oneworld who helped make this a physical book, in particular my editor, Cecilia Stein, who pulled at the text

ACKNOWLEDGEMENTS

further to tauten it, and the copy-editor, Helen Szirtes, for her close and thorough reading of the text.

Viv O'Gara and Katie Maguire-McMorrow, for the lessons on rhythm you didn't know you were giving me.

My ever-respectful children, Alexander and George Conlon, who let me try out the text on them and read a draft version.

And finally Lucy, first sounding board for all my crazy ideas, for all her love and support.